BEAUTY AND THE BRAIN:
THE AESTHETIC COMPASS

All inquirers should be addressed to:

Book Domain LLC.
543 E Louise Dr Phoenix, Az 85050

Ordering Information:
Amount Deals. Special rebates are accessible on the amount bought by corporations, associations, and others. For points of interest, contact the distributor at the address above.

Printed in the United States of America.

ISBN-13	Paperback	978-1-964100-01-2
	eBook	978-1-964100-00-5

Library of Congress Control Number: 2024905207

THE AESTHETC COMPASS

WHAT GALAXIES, TORNADOES AND SNAIL SHELLS HAVE IN COMMON

NeuroAesthetics:
Where Consciousness and the
Physics of the Universe Meet

How do we perceive beauty?

ROBERT W. THATCHER, PH.D.

BOOK DOMAIN LLC
Publish to Perfection

ACKNOWLEGEMENTS

Special thanks to Eric Schwartz for his penetrating insights into sensory to cortical mappings and the Golden proportion. Also special thanks to E. Roy John who supported Eric Schwartz and my abstract explorations of the brain while we were faculty members in the Department of Psychiatry at NYU School of Medicine where many of the ideas expressed in this book germinated. Last but not least, I want to thank all the members of the Neuroguide EEG internet forum for their feedback to improve the science and clinical applications. I want to express a special thank s to Dr. Alexei Berd for his access to an abundance of scientific knowledge from which we all benefit. I wish to thank Dr. Donald Pettit, Alfredo Toro, and George Chaiken for their inspirational contributions and germinal ideas supplied during our many conversations. Lastly, I especially acknowledge the editorial assistance of Dr. Rebecca McAlaster.

CONTENTS

1

CHAPTER

THE STORY OF AESTHETIC FEELING

"In every man's heart there is a secret nerve that answers to the vibrations of beauty."

Christopher Morley

The human feeling of beauty transcends words because its origin is subconscious. The conscious "aesthetic moment" begins as both an awareness and a feeling of beauty. Among the many questions are: Why is a sunset or a child's smile universally beautiful? Why are aesthetic feelings a driving force in the world economy? Why are cost/benefit choices often based on simplicity? Why does beauty to the mind's eye help us decide on lifestyle issues? Why is aesthetic feeling a driving force behind human creativity of architecture, engineering, mathematics, athletics and art?

In this book the argument is made that the answers lie in the unconscious matching of the anatomical structure of the human brain to "Ideal" mathematical forms in the Universe itself, similar to Plato's philosophy of a separate mathematical ideal Universe that humans can only approximate. Plato's ancient hypothesis has recently been tested and confirmed when it comes to the human brain. For example modern Neuroscience discovered that the mapping of the retina (a disk) to the visual cortex (a rectangle) is a logarithmic spiral the same as a snail shell, a Galaxy or a Tornado (Schwartz, 1977a; 1977b; 1980). The human cochlea is another example of this spiral form that maps sound to the cortex and sensory mapping of the skin via a spiral mapping to a straight line in the cortex. Why is there a common logarithmic spiral mapping of our senses? What evolutionary advantage is there to a logarithmic spiral sensory-cortical map and aesthetics?

Also, why is the mathematical logarithmic spiral form of a snail shell, a spiral galaxy, and a hurricane the same logarithmic spiral mapping of the retina to the cortex, the skin and the human cochlea? There must be a deeper meaning. One goal of this book is to explore the deeper meaning of the Platonic ideal mathematical Universe, and the human brain's approximation to this ideal with simplicity and minimal energy forms guided by the aesthetic feeling arising from brain networks. A search of the deeper meaning of the match between brain anatomical mappings and the spiral mappings of the Universe is where this book begins.

First, let us begin with the hypothesis that modern neuroscience measurements of the "aesthetic moment" begins with sensory input (visual, touch, sound) and an approximate 200 ms -500 ms delay before the sensory cortex measures the aesthetic impact. This is an approximate match to a universal Golden Proportion spiral form of brain anatomy (like a snail shell) followed by human frontal-limbic loops which give rise to the feeling of aesthetic appreciation and iterations of thoughts and memories. Next, let us test this hypothesis, based on published science, showing a universally beautiful spiral mapping of the retina and cochlea and skin to the human cortex which is fundamentally involved in high level processing at each moment of time just by virtue of the spatial mappings themselves. In short, this book is about a subconscious phylogenetic force which begins as an immediate aesthetic activation produced by "log spiral sensory maps" in about 100 ms to 200 milliseconds. This is later followed (200 ms to 1 sec) by cortical-limbic iterations for the deeper emotion of the aesthetic context of our perceptions. This all happens in a continuous sequence of brief frames of time that constitute the "spacious present" (Thatcher and John, 1977; Thatcher, 1977; 2016). The proposition is that human consciousness is a phylogenetic force giving rise to an immediate aesthetic feeling by cortical loops and iterations for the deeper emotion of the aesthetic context of our perceptions.

To experience the shared ground truth of beauty, pause and close your eyes and take a deep breath and imagine a **Beautiful Sunset or a Beautiful Flower**. If you were successful and sustained the image of a sunset or flower, then you shared an aesthetic moment of the feeling of beauty that is universal in all cultures and societies. A sensory aesthetic feeling when we view a beautiful landscape, a child's smiling face or a bright flower, is fleeting and may last only seconds, to be replaced by the next moment of our conscious frames of time. The sensory aesthetic feeling is the content of one or more fleeting millisecond frames of awareness that move like a traveling wave in loops of the brain. These biological processes are the pleasure of the aesthetic feeling moment by moment. These processes are elicited by a remarkable and amazing set of universal forms of simplicity and harmony that are correlated with aesthetic moments. The log spiral in physics is also a "minimal energy" form and it is remarkable that the same geometric forms also produce pleasure upon seeing them (e.g. sunset, snail shell, flowers, galaxies, etc). This commonality between the mathematical physics of the Universe, as represented by pure mathematical forms such as the logarithmic spiral, the Golden proportion, and the anatomy of the human brain, remarkably produces moments of appreciation, awe or joy when experiencing beauty. Aesthetic feeling also has evolutionary survival value which is certainly not trivial, and, in addition, reflects a deep and profound truth.

The aesthetic moment can cause one to be breathless or to be "stunned" for a brief moment, which creates a lasting memory that may significantly change one's life path. Examples are, the sudden phase shift of consciousness through simplicity like a sunrise, a mountain top, waves on a beach, a tornado, a mathematical equation, or a correct moral judgment that catches our attention. Then, more profound aesthetic feeling follows the initial minimal energy sensory input which gives rise to aesthetic appreciation of the context and meaning of the sensory input.

Science shows that the aesthetic moment is a two step process. First minimal energy forms imping on our senses that are logarithmically mapped to the sensory cortex. These are followed by unique human iterative loops between the cortex and the limbic system that gives rise to the aesthetic moment. The aesthetic moment can be profound, and it can influence our decisions and directions in life. This I refer to as the "Aesthetic Compass".

The aesthetic moment is also evoked by the recognition of "truth" and the moral intuitive of "what is right". Aesthetic judgment is a moral force in all of our lives. Phrases such as "do what is right" and "truth" are expressions and manifestations of the "Aesthetic Compass". Additionally, the aesthetic compass is also a subconscious force related to the feeling of "justice" vs "injustice", a unique property of the phylogenetically evolved human brain that is absent or minimal in non-human primates.

Beauty is a unique and special human feeling that comes effortlessly and is a minimal energy form in the brain. For example, lets take a deep breath and imagine a sunset or flower or the beauty of an image of a galaxy in deep space. Like water flowing down a hill, sub-conscious anatomical ideal mappings follow the ideal mathematical "Golden Ratio" or "Divine Proportion" (i.e., 1.618033.... Huntley, 1970) and are the essence of simplicity and beauty, the beginning of the aesthetic moment.

To Summarize, the human cochlea is a logarithmic spiral. The retina has a logarithmic mapping to the visual cortex. A straight line in the somatosensory cortex maps as spirals winding around our limbs (Schwartz, 1977a; 1977b). Logarithmic spirals are Golden Proportions which represent a common minimal energy format for basic human perception as the genesis of the aesthetic moment. Figure 1 are examples of the Golden Ratio in nature which is immediately aesthetically pleasing to the eye and is an anatomical mapping of sensory receptors to the human cortex. The Golden ratio's immediacy and simplicity draws our attention, inspires reflection and thoughts, and changes in the trajectories of our life.

Line Segments in the Golden Ratio

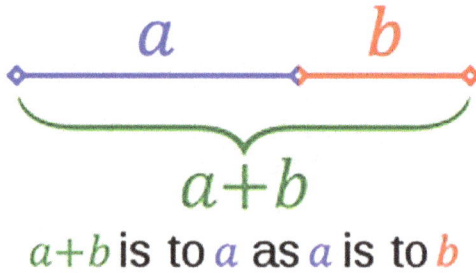

$$a \qquad b$$

$$a+b$$

$a+b$ is to a as a is to b

Logarithmic Spiral and the Fibonacci Series

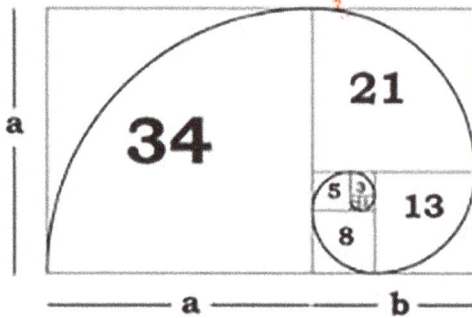

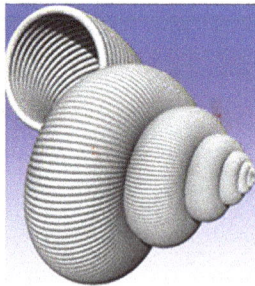

*Fig. 1-The **Golden Ratio** is a special number found by dividing a line into two parts so that the longer part divided by the smaller part is also equal to the whole length divided by the longer part. It is often symbolized using phi, after the 21st letter of the Greek alphabet.*

It is argued in this book that our subconscious aesthetic compass stimulates actions and gives direction. For example, only humans travel to the sea shore just to see a sunset! No one that I know of has witnessed a group of dogs or cats or monkeys walking to the sea shore just to see a sunset! A feeling of beauty is immediately relaxing and is a positive feeling mixed with awe. The power of the feeling of beauty is when the beauty of the moment like a sunset or flower comes to consciousness. Then, human beings pause and take it in with an urge to prolong the experience. The feeling of beauty involves brain neuromodulators like dopamine and serotonin that modify synapses, and thus is a reinforcing, positive feeling that acts like an "aesthetic compass" in that brief moment which gives rise to changes in behavior and the direction of future life experiences.

Non-human primates lack the phylogenetic development of the human frontal lobes to give rise to high impact aesthetic appreciation and emotional depth of feeling. Aesthetics starts as a fundamental subconscious force based on minimal energy sensory matching of the Golden Proportion logarithmic spiral form of the human brain as an initial information format. It then maps to frontal lobe networks of the brain that are essential for creativity, and survival, self-awareness and iterative loops of thoughts that strive to understand the past, future, and to imagine worlds beyond. Much of this book is dedicated to understanding the science behind the aesthetic feeling by postulating the existence of iterative "frames of time" which are matched and mismatched with new experiences to give rise to feelings of novelty at those moments of time that also share the iterative nature of "self-similarity", "fractals" and the "Golden Proportion". But there is an even deeper level and I will strive to show this in how the aesthetic feeling involves a fundamental linkage between the physics of "minimal energy" in the Universe and mechanisms of uncertainty reduction inside the brain which give rise to the appreciation of aesthetics. I believe that Plato was one of the first to explain this deeper level of aesthetics when he envisioned an ideal mathematical universe (Agnati et al, 2007; Cooper and Hutchinson, 1997). Roger Penrose (2005), a Nobel mathematician, in his book "The Road to Reality: A Complete Guide to the Laws of the Universe" is a modern continuation of Plato's early intuition. These and similar works form the scientific foundations of the Neuroscience that I rely upon to help link aesthetic feeling to the brain.

The mystery of the feeling of beauty raises questions that are fun to explore. For example : 1-What is the biological survival value of the feeling of beauty? and 2-Why is the feeling of beauty so rewarding, and acting as a magnetic compass that attracts re-experiencing it? A starting point in answering these questions is to begin with understanding the relationship between the brain and the physics of the Universe that contain the same atoms that the brain is made. For the brain obeys the same laws of physics as does all material known to man. By this, I mean today's 2022 accepted mathematical physics that constitutes Quantum Mechanics and Einstein's special and General Theory of Relativity and the other laws of the Universe. These purely ideal applied mathematical forms give rise to every-day comforts and technology such as the ability to travel at high speeds, the TV, computers, space travel and also, unfortunately, destructive like atomic bombs, etc.

How is it that the human brain feels beauty yet can also create atomic bombs? Astoundingly, the relationship is by virtue of the beauty of the ideal perfect form, driven by a Universe of ideal perfect forms, which, as Plato argued, exist as a separate ideal Universe each time we match and mismatch. Then only approximate the ideal form at each moment of time in our everyday moment-to-moment life. Ironically, the Platonic ideal of applied mathematics in an atomic bomb also served as an aesthetic compass that guided concentration and skill to perfect the atomic bomb. The equations were guided by the same feeling of aesthetic perfection as a University Professor or engineer creating new inventions.

Plato's simple mathematical forms, such as a circle, a triangle, a spiral, a rectangle, radial lines and many other pure mathematical forms have been considered "beautiful" for billions of people for centuries of time . An amazing fact is that the human brain's sensory mappings share the same Platonic fundamental forms! As mentioned previously, the human cochlea is a logarithmic spiral form, like a snail shell! Why is this? The retina is a disk and the mapping to the visual cortex is a logarithmic spiral, like a snail shell! Why is this? Why does the cochlea have the same form as a Galaxy or Tornado? A straight line in the sensory motor cortex maps as a spiral on the limbs! Why is there the same common mapping as for sound and sight and touch? Why is there a common mathematical physics format of the human brain of a logarithm that effortlessly creates multiplication by summation for sensory input?

1.1 COMPUTATIONAL NEUROANATOMY

The answer? Dendritic synaptic summation is actually multiplication along the dendritic axis of neurons because logarithms are multiplication by addition. This fundamental fact, as described by Eric Schwarts (1977a; 1977b; 1980), results in computational neuroanatomy that gives rise to minimal energy matching of the mappings of the external world to the physics of the brain's mathematical forms at each moment of time! The brain's job is reducing uncertainty in an uncertain universe, because our anxiety increases with uncertainty. Our ability to predict the future, or our ability to meet expectations or adjust for failures is necessary for our survival. The brain uses instantaneous computations of match and mismatch of form and spatial mappings in milliseconds of time to predict the future. It takes action by comparing match-mismatch of the success of actions in instantaneous steps toward minimal energy states and aesthetic forms that are the best fit at each moment of time.

Information flow from the sensory surfaces such as the retina to the visual cortex and nearby re-mappings requires about 250 ms, and then another 250 -500 ms to the frontal lobes, anf finally loops to the limbic system to evaluate the value of a stimulus in about 1,000 to 1,500 ms (Thatcher and John, 1977; Thatcher, 2016). It is postulated here that the information mappings when the match approximates the golden proportion is maximally efficient with less effort than other ratios and forms. Hence synchrony and resonance of larger numbers of neurons occurs. I will attempt to explain that the Neuroscience of aesthetics concludes that the feeling of beauty is a brief moment of anxiety reduction by virtue of the fact it that starts in stage one as a high speed approximation in the direction of a perfect match to a perfect world. The "perfect world" is external to ourselves. It is the Platonic concept of ideal forms that today can be hypothesized to be related to the fact that we only know about 5% of the forces and dynamics of the universe but 95% is unknown to us. The "unknown "Dark Energy" and "Dark Matter". The "Dark energy" is responsible for the expansion of the universe. Dark matter are "black holes" that suck up energy and matter. The brain, capable of knowing the 5% of the Universe is guessing about the nature of the unknown 95%. This accepted "fact" of modern 2022 science is truly astounding!

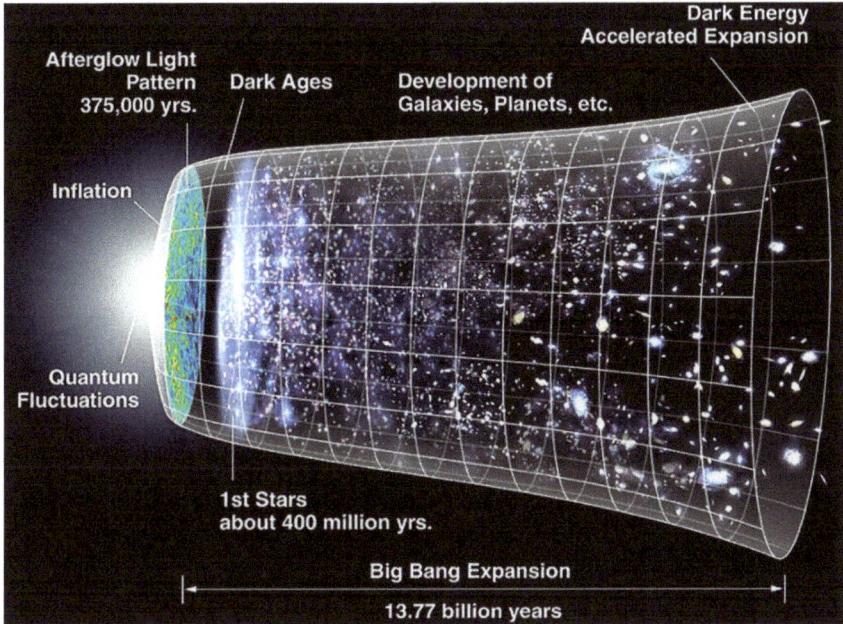

The beauty of seeing a newborn baby is much more complex than the starting point of logarithmic mathematical forms or understanding the relationship between the brain and the feeling of beauty. This feeling can be wonderful but dependent on phylogenetic parental emotions and family history. Mathematics of aesthetic feeling is transformational and based on numbers and is universal. It is best measured independent of personal associations. The main focus of this book is answering the deeper and more profound questions about human consciousness, such as, why is the feeling of beauty influenced by logarithmic spirals in space and/or time? Why is the feeling of beauty associated with the Golden Proportion? Why do the sensory mappings of the human brain follow the architecture of the same Golden Proportion form as painted by Leonardo Devinchi and other artists? Why does the immediate sensory awareness of beauty cause the feeling of relaxation?

The answer is that biologically based logarithmic mathematical mappings in the human brain and logarithmic mathematical forms in the external world are perceived as beautiful. An example is sensory-cortical mappings and the golden proportion and the logarithmic spiral. Is there a relationship between the feeling of beauty in music and the spiral form of the human cochlea? Why is the mapping of the retina to the visual cortex also a logarithmic spiral map? Why does a straight line of sensory cortical neurons map as a spiral to the limbs like a "roman soldier's" boot? Why is there a common format for sensory mappings of the human brain? Why are mathematical forms such as spirals, triangles, circles, radial lines, symmetry, harmony and many other mathematical art forms intrinsic to sensory cortical maps? The second stage of aesthetic feeling is when the brain produces predictions of the future in about 100 ms to 200 ms windows of time that are then subconsciously matched and mismatched to "anatomical reality". The third stage at about 250 ms to 1,500 ms., involves frontal-limbic system loops that reduce uncertainty in frames of time that iterate to expand understanding and reduce uncertainty and constitute the content of our consciousness. The release of reward neuromodulators, such as dopamine, give rise to synaptic modifications in limbic and cortical brain networks to reduce anxiety and guide the "aesthetic compass" (see Chapter 2 on Neuroaesthetics).

To summarize, the Neuroscience definition of aesthetics involves three stages: 1-sensory information engaging the senses via log spiral sensory to cortex maps (about 50 ms to 100 ms), 2-match-mismatch and predictions of the future (about 100 ms to 300 ms) involving reduced stress at its core with increased sympthetic activity, reduced amygdala neuron activity and activation of the pleasure network (Brown et al, 2011; Berridge and Kringelbach, 2015). Reduced amygdala reactivity is important because the amygdala neurons are like the "tea party" in American history, a small group that shouts loudly and shifts the dynamics of a population of 100 billion neurons in a few milliseconds. TV advertisements, the internet, TV, radio and print media are input that is unconsciously sampled at short intervals of time. Stage three (300 ms to 1,000 ms) is when conscious awareness receives information for brief frames of time that are pieced together to form the stable continuous world that all conscious individuals experience.

1.2 WHAT ARE TRUTH AND BEAUTY

The feeling of "Truth" drives us hither and thither over the course of our lives like nodal points in the stream of life's journey (anonymous).

"Our brain is constantly trying to find meaning in associations, connections, data and patterns. We are trying to make it out of the information quantity that is being presented. We also try to connect new information to our past experiences and knowledge stored in our minds. When we find a pattern that is meaningful to us, we add it to our perceptual map. If it connects to the knowledge already stored in our minds, we learn. When we can make those connections, we get a sense of relief from the anxiety,

confusion or stress that accompanies data, facts and figures. " [Rabinovich et al, 2012, p. 6].

"Ideal mathematical forms are not revealed directly and the brain is a massive and highly organized biological process of match mismatch of expectations and predictions with dynamical error correction for uncertainty reduction by quantum and classical physics that approximate ideal forms at each moment of time." (Thatcher, 2016, Chapter 9.1).

The feeling of truth and beauty is produced in the human brain and one can ask the simple questions "Why"? and "How"? do such remarkable feelings guide masses of people on earth in their dreams and quests throughout their life journey? Also why and how are the words "Truth" and "Beauty" often synonymous? The answer to this, requires the use of logic generated by the brain. The brain consumes 20% to 40% of blood glucose, weighs only about 2.5 pounds, but creates consciousness capable of reflection and asking the questions in the first place. This is at the heart of Godel's mathematical proof of self-referential truth. It is also centered on Einstein's General Theory of Relativity which is itself a mathematical truth based on a Platonic Universe of "ideal" forms, for example, the ideal square, the ideal triangle, Pythagoras theorem, etc.

The linkage between brain and aesthetic feeling starts with questions like: Why does that two and half pounds of brain tissue consume 20% to 40% of blood glucose? What work does this disproportionate amount of energy do? How does this brain produce consciousness, thoughts, feelings and goal directed movements? Fortunately, we know the general answer to these questions and it is "Electricity". The last 70 years of neuroscience has established that most of the brain's metabolic energy is utilized to create electricity in neurons connected together in loops. This energy is activated by summated synaptic potentials that produce digital action potentials which send signals along fibers that connect to each element in the loop with branches to other loops. Some of the loops are inhibitory local loops and others are long distance excitatory loops that operate by a balance between inhibition and excitation. Oscillations in loops and membranes result in the scalp recorded electroencephalogram or EEG, which is a few microvolts at the scalp surface but ranges from a few milivolts to a volt inside the skull (Thatcher and John, 1977; Thatcher, 2016). Due to low skull conductance, scalp EEG is measured in millionths of a volt. It is remarkable but fortunate that even with such low voltages the EEG nevertheless allows for precise and reliable measures of phase shifts and phase locking (synchrony) of functional hubs and modules of the brain in milliseconds of time and about a centimeter of spatial localization (Thatcher et al, 2008; 2009; Thatcher, 2016).

The brain is not like a computer. It depends on flows of information in regulated loops, not inflexible and fixed circuits. Unlike a computer the brain creates the future by match-mismatch of its actions. It does this in the face of uncertainty and having to learn the procedures on the fly. The process is like consciousness in pursuit of itself, always being one step behind, thereby being guided by momentary instabilities that minimize negative outcomes and maximize positive outcomes in a continual sequence of metastable brain states. The least effort principle of physics guides this process like water flowing down a hill, a least effort state represented by the synchrony of masses of neurons, only a few milliseconds before one acts. As quoted by Rabinovic et al, (2012, p. 2) "Life is like playing a violin in a concert while learning to play and creating the score as you are playing." Desire and urges result in an action that matches the challenges of an estimation of reality of a moment. A final decision occurs by the selection and temporary synchronous binding of billions of neurons.

One of the iterated themes in this book is the top-down and bottom-up information flow guided by least-effort physics to achieve momentary goals and end points to be replaced by the next temporal frame and minimal energy state, which again matches and mismatches with expectations in a continuous sequence of metastable states. The physiological process of recruiting neurons is a brief 20 ms. to 80 ms. phase shift analogous to a "shout-out, who's available?". The recruited neurons are then phase locked for about 200 ms. to 500 ms. (Thatcher et al, 2009). This process is the calculus of minimal energy states that give rise to sequences of metastable brain states that reduce uncertainty to the extent possible at each moment of time (Pikovsky et al, 2003; Tass, 2007; Thatcher et al, 2008; 2009). All of these processes are measurable in the quantitative electroencephalogram (QEEG) and magnetoencephalogram (MEG) where clinical interpretation is via the linkage of symptoms and complaints to dysregulation in localized brain regions and networks. EEG is unique in that it inexpensively measures real-time changes in homeostatic neuroplasticity by comparisons to a reference database where synapses and brain networks are modifiable by biofeedback (Hellyer et al, 2015).

Information flow is in the form of coexisting and recursive re-mappings from primary sensory regions to higher order systems and bottom-up and top-down regulatory control that results in adaptive homeostasis of feelings, perceptions, actions and consciousness in successive frames of time. Figure 2 illustrates the information flow nature of top-down and bottom-up sensory flows and loops between the limbic system and neocortex. The basis of motivation and drives and visual cortex to frontal lobes (bottom-up), and frontal to visual (top-down) flows of information in continuous match and mismatch of expectations and reactions to novelty. Minimal energy selection occurs via a phase shift and phase lock as the neurophysiological binding or specification where large masses of neurons are momentarily cross-frequency synchronized. The synchronization is within large collections of closely interconnected neurons referred to as Hubs. Higher degrees of connectivity between Hubs is referred to as modules. The highest degree of interconnectivity is called the "Rich Club" in the Human connectome literature (Sporns, 2013). The network and "small-world" organization of the human brain are another manifestation of the physics of minimal energy flows and the efficiency of reducing uncertainty (Thatcher et al, 2016).

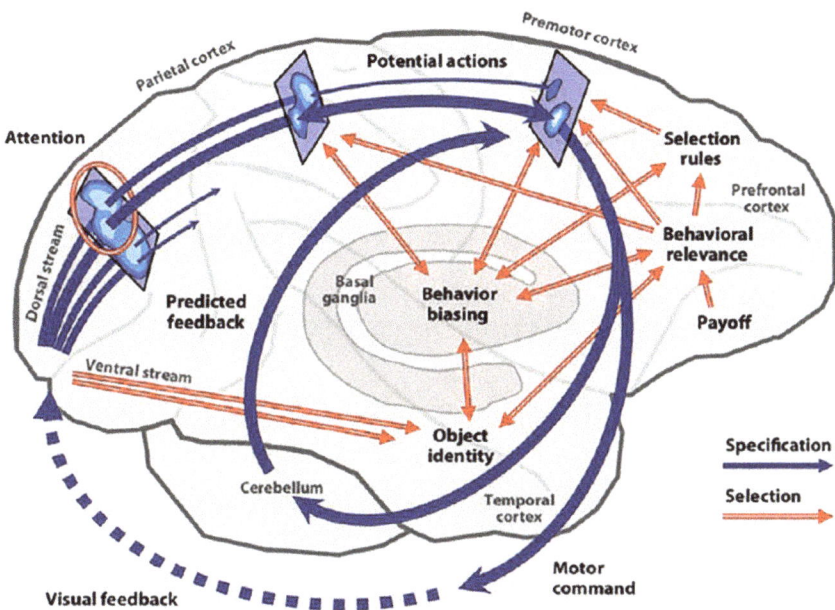

Fig. 2. Illustration of brain information flow that can only be measured by the electroencephalogram using computers. Non-qEEG or visual examination of complicated EEG traces without quantification is incapable of identifying the millisecond dynamics of the human cerebral cortex. Information flow is from primary sensory systems toward the frontal lobes and from frontal motor regions to motor commands as represented by the Blue arrows. Selection of loops of synchronous neurons to mediate adaptive behavior occurs during brief temporal frames as represented by the Red arrows. Phase shift is represented by the red arrows and phase lock is represented by the blue arrows. This figure and the concepts of phase shift and phase lock in the coordination of large masses of neurons in functional modules and hubs is what gives rise to the human electroencephalogram (Thatcher et al, 2008; 2009a; 2009b). EEG biofeedback and the reward system represented as "pay-off" is mediated by dopamine that modifies synapses and is the reinforced and operantly conditioned neural event. From Rabinovich et al, 2012. (See chapter 6 for details).

1.3 PLATONIC MATHEMATICAL UNIVERSE

"... no examination of Einste in's brain has ever shed much light on the sources of his genius, a light as great as the mysteries of the universe that it penetrated." (pg. 92 -"Albert Einste in: The Enduring Legacy of a Modern Genius", Time Magazine, 2014).

"There is also an undoubted deep mystery in how it can come to pass that approximately organized physical material can somehow conjure up the mental quality of conscious awareness" (Roger Penrose, "The Road to Reality: A complete guide to the laws of the Universe", 2005, pg 21).

"There is also a mystery about how it is that we perceive mathematical truth. It is not just that our brains are programmed to 'calculate' in reliable ways. There is something much more profound than that in the insights that even the humblest among us possess ...". (Roger Penrose, "The Road to Reality: A complete guide to the laws of the Universe", 2005, pg 21).

"Consciousness in pursuit of itself is always one step behind" (Thatcher and John, 1977).

UNCERTAINTY REDUCTION AND TRUTH AND BEAUTY

Consciousness is defined as a natural evolution executive process that vetoes or permits the commitment to action. It operates as a delayed 'subjective viewer' of serialized frames or moments in the stream of the content of consciousness of subjective feelings and thoughts, feelings of fear, hope, happiness, sorrow, awareness of something new, pain, pleasure, etc. Consciousness, like a CEO, operates at the top of a vertical phylogenetic hierarchy from brainstem to neocortex of preconscious processes that precede conscious awareness. There are time gaps between the occurrence of events and the awareness of the event which vary depending on the rhythms of the brain at the time of the event. For example, during sleep the sensory relays are inhibited, EEG exhibits large slow waves 1 -3 Hz, and there is no or minimal conscious awareness of changes in the environment when the brain is in the sleep state.

Historically, theories of consciousness are divided into two general groups: 1-'Mind or Consciousness' as separate from the atoms and neurons that make up the brain (so-called mind vs body dichotomy) and, 2-Consciousness as an emergent property of natural selection by which a delayed executive has survival value. As a self-reflective process, consciousness is a physical property of atoms, 'emerging' from the collective actions of atoms and subatomic particles at classical and quantum mechanical levels. Centuries of thought have concentrated on life, the stars, the past and present and the nature of the human mind. The nature of consciousness and free will and the contradictions of physics is a significant challenge for the approximate 2 -3 pounds of mass, called the brain, to comprehend. This is why it has taken over 100,000 years for the information limited by a small brain operating in about 200 ms time frames to finally externalize words, thought and ideas and to learn how to efficiently store and transfer knowledge from one generation to the next. Slowly, over time, logic and mathematics and the scientific process of hypothesis testing, gave rise to the enormous mathematical and physics accomplishments of the 1800s -1900s, in which the laws of the universe can be written on half of a page of paper (Richard Feynman et al, 1963). In the 21st century after huge effort and expense physics has demonstrated the existence of "dark matter" (e.g, contraction of matter into neutron stars) and "dark energy" (opposite of gravity, growth/expansion) that constitute about 95% of the universe. This means that conscious awareness, at a given moment of time is at most only aware of about 5% of the universe. Therefore consciousness must be an emergent property of a yet unknown dynamic. Quantum Mechanics, in which the Higgs Boson has been discovered, confirmed that Gravity is a particle, (i.e the so called "God" particle) thereby confirming the Standard Model of quantum mechanics, which itself is a part of the dynamic between dark energy (growth/expansion) and dark matter (contraction). Recent telescopic measures from the South Pole have confirmed by experiment the expected distribution of light at the beginning of the universe due to Gravity, consistent with Einstein's "General Theory of Relativity". This indicates that a very small particle caused the expansion of the universe, as a near instantaneous explosion spreading out from a point source creating a uniform fabric of the universe as described by Einstein's "General Theory of Relativity" as well as "Quantum Mechanics". Consciousness must therefore be part of the fabric of the universe in the context of the entropy of the universe as "borrowed time" in the "warp" of the fabric of Einstein's "space-time" for short periods of quantum mechanical time, like waves on the surface of a vast ocean. According to the General Theory of Relativity there is no net loss or gain of changes or warping of the fabric of space-time, only transforms from moment to moment. Based on propofol anesthesia studies, consciousness appears to represent a nonlinear dynamical separation of the vertical coupling of the reptilian brain to the human cortical primate brain (et al, 2010). Loss of consciousness is a "quenching" of synchrony between the cortex and brain stem, in which sleep and unconscious states are characterized by a space-time collapse of the lower and higher frequencies of billions of neurons that are rapidly transformed into extreme hypersychrony at a lower dimensional state or the unconscious state (Breshearsa et al, 2010). Awakening and conscious awareness is a reversal of the collapse of hypersynchrony into a dancing dynamic of brief metastable states of cross-frequency separation between the neocortex and the brainstem. The dance between the primitive dinosaur brain and the human neocortex reduces uncertainty and is necessary for survival. This dance is not trivial because it is quantum mechanical and classical in the fullest meaning of these terms.

Roger Penrose in his book "The Road to Reality: A complete guide to the laws of the Universe" (2005) explains how reality is actually a separate mathematical Universe similar to what Plato conceived of as ideal forms. Earlier, Penrose (1994), developed an elaborate science of consciousness using the mathematics and physics of quantum mechanics and relativity theory which also relies upon the ideal mathematical Universe of Plato. Today, human beings know only about 5% of the physical Universe, although experiment and mathematics allow estimations about the nature of the remaining 95% that is outside the range of our senses and conscious awareness. Ideal mathematical forms are not revealed directly. As a result, the brain is a massive and highly organized biological process of match and mismatch of expectations and predictions with dynamical error correction. This reduces uncertainty by approximating ideal forms at each moment of time. Penrose's quantum mechanical model of consciousness is a fascinating read. It is consistent with modern quantum mechanics since "warm and wet" quantum processes were discovered (Panitchayangkoon et al, 2010; Engle et al, 2007). An extension of Penrose's model is to drill down to the concept of a Planck constant of consciousness. Planck's constant is based on jumps or nonlinearities in the behavior of atoms discovered in the late 1800s. Max Planck discovered a minimal time frame between classical physics and quantum mechanics, a quantum entity called the "photon".

The first link to quantum mechanics and the brain is the fact that the brain is mostly made of water, i.e., H_2O. The 2nd link to quantum mechanics is that the density of high energy protein/lipids of the human brain, contained in a small space with van der Walls quantum mechanical forces operating, results in time warp quantum reality instantly between dark energy and dark matter in every atom of every human being. A bridging concept between quantum mechanics and the brain is a type of Planck constant of consciousness based on measured frames of time of perception and consciousness with masses of energetic molecules and atoms consuming about 20% to 40% of blood glucose and concentrated in a small bony skull and weighing only about 2.5 pounds. Let us briefly explore the physics of the brain. Lets start with the conservative estimate that the 1 killogram brain (2.2 pounds) is made up of 100 billion neurons or $1 \times 10 \ 12$ with 1×104 synapses and 1×106 ionic channels $= 1 \times 1022$. Avagrado's number is 6.7×1023 as a measure of the atoms per unit volume that relates classical physics to the atomic level of atoms. Conscious awareness occurs in discrete "time frames". The large density of energetic atoms that constitute the brain necessarily involve quantum mechanical van der Walls forces. For example, a hydrogen atom is electrically neutral but has a spatial dipole distribution at the quantum mechanical level. It is the van der Wall's quantum level dipole distribution of electrons that gives rise to the remarkable properties of water, which constitues about 85% of the human brain. Also, it is worth noting that it is the shape of an H_2O water molecule is a minimal energy atomic form that dynamically oscillates and approximates a mathematical ideal form which gives rise to the water molecule (Plato and the subconscious Aesthetic Compass). In any material, there is a constant "battle" between the bonds and the kinetic energy of the individual molecules -the bonds try to keep the molecules together while the kinetic energy tries to separate them. In a solid, the bonds are strong enough to keep the molecules firmly in place, connected to one another, and to overcome the kinetic energy. In a liquid, these bonds are weaker, and are constantly being broken and reformed as the molecules move around -not strong enough to keep them in place, but strong enough to keep them from flying away. The brain is mostly water and protein/lipids operating in a very dense highly energetic space. It is the integration of macro and micro graph theory that links classical and quantum mechanics and the brain.

1.4 LIFE, AESTHETICS AND MINIMAL ENERGY

In the 1920s the relationship between aesthetic feelings and "complexity" was redefined by Birkhoffin in terms of "perceptual effort". This linkage between the complexity of an object and perceptual effort is paralleled by a similar linkage between mathematics and the simplicity of the physical laws of the universe. For example, the complex logarithm function is characteristic of the global and local structure of the sensory mappings of the brain. This logarithm function may also be used to describe the pattern of electric or magnetic fields, the velocity flow of a fluid, or the distribution of a diffusing chemical reactant. The basic developmental reason for this commonality is that structures of these sorts require minimal encoding. That is, they represent the most parsimonious and economical methods of controlling dynamic flow, and in the case of living matter, the growth of form. The unifying and simple nature of these observations indicates that their commonality is ultimately an expression of the "Variational Principle" or the principle of "Least Effort" in physics. This principle, as reflected in the calculus of variations, is a description of the processes by which nature finds the path of least resistance, or a solution of elegance, simplicity and parsimony in the resolution of conflicting forces and evolution in nature (a sunset, a flower, a tornado). The mechanisms of aesthetics and perhaps more generally, perception, may directly involve a similar linkage. In this case between simplicity and perceptual effort, in which the least effort expression for the growth of living forms (the Golden Proportion) is matched by the least effort expression for the evolution of the physical universe. In other words, a fundamental aspect of the human aesthetic feeling involves a match between the organizational laws of the atoms of the brain and the organizational laws of the atoms of the environment or space external to the brain. The mathematical variational principle of Euler-Lagrange and Hamilton is a universal expression that applies to living matter and human consciousness. In the case of human consciousness the negative entropy of synchronous discharge of millions of neurons time linked to the present, and matched and mismatched to memory and expectations of the future in 80 to 300 millisecond intervals of time are sequential minimal energy states.

Finally, we must ask: what is the linkage between "living" and "non-living" matter, the "variational principle" and the "golden proportion"? It is reasonable to assume that this link arises from the operation of templates in nature, where the duplication of a basic form is the simplest and most economical method for survival and growth. An example of the role of templates in inanimate matter is the hydrogen atom whose basic structure is duplicated to create helium, and so on until all of the elements of the periodic chart exist. In the case of "living" matter, the way in which elements are organized to create a whole is of fundamental importance. The temporal and spatial coordination of parts in the context of the whole is essential for living systems. The old aphorism that the "whole is greater than the sum of its parts" is an example of this fundamental principle of life.

The golden proportion is of special relevance in the organization of life by virtue of the fact that it is the only proportion in which "the ratio between the greater and the smaller part is equal to the ratio between the whole and the greater part". That is, there is no other ratio in which the larger and smaller parts are related to each other by virtue of their relationship to the whole. Duplication, when operating on the golden section or proportion, results in the property of gnomic growth in which the relationship of the parts to the whole is preserved, independent of size. Thus, through the evolutionary process, nature operates on those "minimal energy forms" which survive, by simply duplicating them (sometimes with ratio variations) over and over again. The golden section provides for a numerical proportionality by which complex growth can economically occur. This economy is due in part to the simplicity of multiplication by addition., i.e., the "Golden Proportion" itself.

1.5 SELF-SIMILARITY, RHYTHM, SIMPLICITY, PROPORTION

Aesthetics is defined by the Webster dictionary as "The branch of philosophy that provides a theory of the beautiful and of the fine arts". In psychology, aesthetics is a branch of perception concerned with the feelings of the appreciation of beauty which comes effortlessly and immediately upon contact with certain objects and sounds. In neuroscience, aesthetics is sensory mappings that are minimal energy. There are many theories of aesthetics, but all such theories subscribe to a common set of geometric and temporal concepts; namely, the concepts of **"rhythm"**, **"harmony"**, **"simplicity"**, **"proportion".** We generally associate the term "rhythm" with the arts concerned with the time dimension (e.g. poetry and music). We associate the notion of "proportion" with the arts concerned with the spatial dimension (e.g. architecture, painting, decorating, etc.). However, as described by Ghyka (1977) this superficial distinction between rhythm and proportion is a recent invention. For example, ancient Egyptian and Greek philosophers did not recognize these distinctions and, instead, considered rhythm as the most general concept, which dominated not only aesthetics, but also psychology and metaphysics. According to Pythagorean doctrine, rhythm and number were unified, with everything in the Universe arranged according to ratios of numbers. Plato formally developed Pythagoras' aesthetics of number into the aesthetics of proportion. For example, the technique by which proportions were linked so as to get the right correlation between the whole and its parts was called by the Greek architects "simplicity of symmetry". The result obtained where this technique was correctly applied was the "eurhythmy" of the design and of the building. For the Greeks, architecture was not only "Frozen Music, but living music" (Ghyka, 1977). In modern times a fundamental process called "Self-Similarity" was discovered to be one of the components of aesthetic form such as in animal and plant biology. Mandelbrot (1982, page 34) characterized self-similarity thus: "When each piece of a shape is geometrically similar to the whole, both the shape and the cascade that generate it are called self-similar."

One of the most influential rhythmic, and simplest proportions was called by Pythagoras the "Golden Proportion" and by Leonardo Devinchi the "Divine Proportion", which is an irrational number called Phi = 1.618033…… As mentioned previously, this is not a trivial number, because it is the only number in the universe whereby a small segment of a line creates the whole or total line segment. Pythagoras proved that the "Golden Proportion" is the simplest continuous proportion. Described mathematically as $a/b = b/c$, or b^2 Fibonacci series $1/3, 3/5, 5/8, \ldots$

Aesthetics is related to the feelings of beauty, simplicity and harmony. The aesthetic feeling is powerful. It determines human decisions and inspires the human mind to create monuments, strive toward ideals, and achieve some of the greatest human accomplishments. The aesthetic feeling is also fragile and is quickly replaced by base emotions such as anger, fear and negative thoughts. A main hypothesis of this book is that there are at least two primary physiological and anatomical components of the human aesthetic feeling: first is Computational Neuroanatomy of the primary sensory systems by virtue of a logarithmic golden proportion mapping from sense organs to the cortex, with delays around 250 ms. Second, is a re-mapping of the primary sensory information to the frontal lobes and default network with a delay of about 1,000 ms which is involved in aesthetic appreciation (Nadal et al, 2008). This hypothesis is supported by the studies of Cela-Conde, et al (2018) that support the existence of two sequential brain networks.

Figures 3 and 4 are examples of MEG signals corresponding to the participants' stimuli appreciation, grouped according to beautiful and not beautiful conditions. The initial network is sensory input. This is followed in time by frontal, temporal and cingulate cortex inputs comprising the "default network". These particular brain networks are sequentially related to the aesthetic experience because they occur in specific frames of time. They give rise to a hypothesized two-compartmental model. Compartment #1 involves sensory input to the primary and secondary sensory cortex in the time frame of 250 to 750 ms. Compartment # 2 is a delayed response in the frontal lobes and default network in the 1,000 ms to 1,500 ms interval which gives rise to aesthetic appreciation.

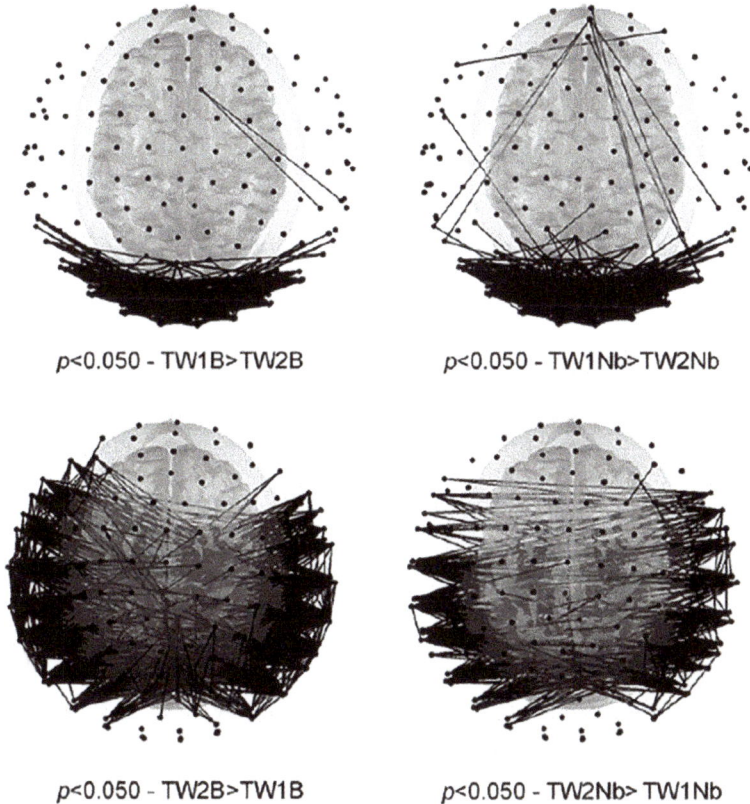

p<0.050 - TW1B>TW2B   p<0.050 - TW1Nb>TW2Nb

p<0.050 - TW2B>TW1B   p<0.050 - TW2Nb> TW1Nb

Fig. 3. MEG signals were split into three time windows and two evaluative conditions. Artifact-free time windows of 500 ms before stimuli projection were manually extracted for further connectivity analysis, constituting time window (TW0). After each stimulus onset, 1,500-ms artifact-free epochs were divided into two additional time windows: TW1, 250–750 ms; and TW2, 1,000–1,500 ms. The length of the windows was determined by taking into account the time span in which brain activity can reach frontal areas during aesthetic appreciation. Before 250 ms, cognitive processes related to aesthetic appreciation rely mostly on retinal-cortex mapping via the Golden Proportion logarithmic spiral for visual-processing. In turn, MEG signals corresponding to the participants' stimuli appreciation were grouped, constituting the beautiful and not beautiful conditions. Synchronization in Time Window 1 (TW1) and Time Window 2 (TW2) under beautiful (Left) and not beautiful (Right) conditions. From Cela-Condea et al, 2018.

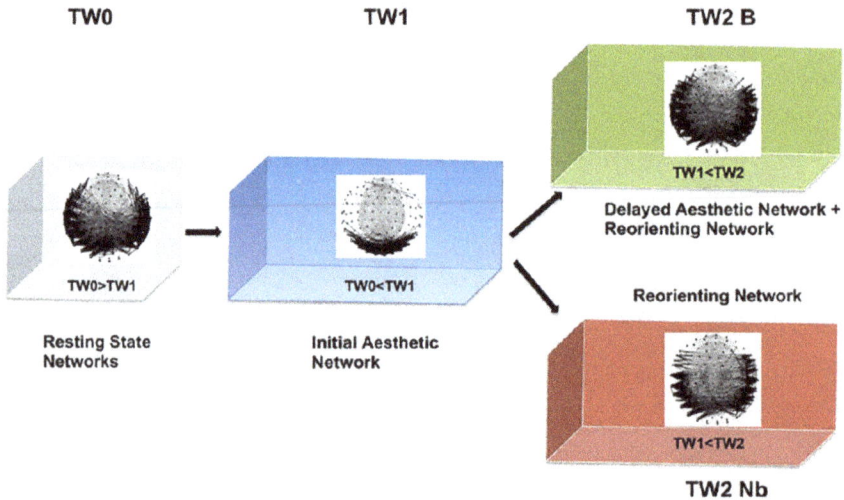

Fig. 4-Dynamics in the appreciation of beauty. TW0 networks (illustrated by the TW0 > TW1 comparison) fade during TW1, being substituted by a similar network shared by beautiful and not beautiful conditions (illustrated by the TW1 > TW0 comparison

under the beautiful condition). During TW2, not beautiful stimuli activate a bilateral reorienting network, whereas beautiful stimuli add the delayed aesthetic network, more medially placed (in each condition, TW2 networks are illustrated by TW1 > TW1 comparison). From Cela-Condea et al, 2018.

There are at least four consistent components of aesthetic feeling as revealed in the history of the study of aesthetics over thousands of years that include: 1-Symmetry, 2-Harmony, 3-Self-Similarity and, 4-Synchrony. Starting with these fundamental hypotheses, one can test the idea that aesthetic feeling is related to the match of golden proportion shapes and forms of brain anatomy, to the mathematical shape and form of external inputs like sunsets, flowers, fireworks, etc (Kaplan, 1987; Jacobsen et al, 2006; Lacey et al, 2011).

The Golden Proportion is an important minimal energy form that is represented in physics, mathematics and neuroscience as the logarithmic spiral mapping of sensory stimuli onto the primary cortical regions of the brain resulting in a "least effort" perception and subsequent neural synchrony. Perception studies (Efron, 1967; 1970a; 1970b; Allan, 1978) have shown that awareness is divided into "perceptual frames" approx. 100 ms to 1,000 ms in duration, and each perceptual frame contains minimal energy forms that occur within that perceptual frame.

I have developed what are called the "Laws of Aesthetics" one of which is where external forms that are maximally simple and maximally matching (matching to minimal energy forms in the brain) yield a maximum aesthetic feeling (dopamine and least stress, least effort). Aesthetic self-organization is hypothesized to involve a "least action" principal of matching between the perceptually organized atoms of the brain and the temporal convolution of minimal energies external to the brain.

Aesthetic feeling acts like a compass called the "Aesthetic Compass" that directs our moments of thought and our decisions. It shapes the paths of our lives at each moment of time. It only occurs after the hierarchy of base needs is met and one is calm and not agitated. The aesthetic feeling is a subtle feeling, one that is blocked by strong negative feelings and comes immediately and is pleasurable. For example, the aesthetic feeling comes effortlessly upon the perception of something "beautiful", such as a flower, a sunset, a star, a wave, a smile, a sound produced by nature or music, etc. Each of these events share the common feeling of "beauty", which initially comes to mind immediately and effortlessly and without conscious reflection or conscious investigation. The primary aesthetic feeling of beauty, however brief in the stream of consciousness, is a transformational feeling. It may have profound and lasting affects on our future actions and future perceptions. It is in this sense that I align the aesthetic feeling with the concept of an "aesthetic compass". The aesthetic compass can be visualized as a small and delicate spinning gyroscope that can be easily perturbed by base emotions (anger and fear) and delicately aligns its spin in calm moments and moments of creativity and pleasure.

As explained in chapter two, I use the word "compass" to indicate pointing toward the ideal mathematical form of "minimal energy" as defined by the "principle of least effort" in physics and the Hamiltonian equation of motion. This is expressed as the golden proportion in which neurons in the medial frontal lobes and the ventral tegmentum dopamineric network becomes active. The aesthetic feeling like all feelings has a neurophysiological origin and dopamine and reward and pleasure networks are inextricably bound and also dependent on cortical resonance via computational neuroanatomy where the Golden Proportion is a transfer function.

The Golden Proportion has an amazing mathematical property as well as a fundamental neurophysiological property that is linked by the physics of the universe, this is described by the area of mathematical physics called variational calculus and "least effort" or "least time" of kinetics and quantum physics (Feynmann 1963). Out of necessity this book will use mathematical equations as a succinct language for those interested in mathematics. For those not proficient in math the equations will be explained and described by words so that it will not be necessary to be proficient in mathematics. Conceptual linkages are important, and math is like a beacon or a flashlight that the human mind uses to explore and understand the darkness and mystery of the universe. Questions like, Where am I?, Who am I and why am I? are questions shared by all human beings for all time. The continual pursuit of the answer to these questions is a driving force in the human psyche, related to curiosity, the drive to explore, to learn and avoid danger, reduce uncertainty and master the environment. At the heart of this pursuit is the "Self" or self identity. This process, as described by Piaget (1975), begins early in life and from it begins our self-narrative and building of elaborate memories of our life's story as we grow older. A network in the brain called the "Default Mode Network" (DMN) is responsible for this self-narration and consolidation of memory and our life experiences. The default network or DMN is turned off when one attends to events external to self (Fox, 2005; Thatcher, 2016). Self-reference is also a fundamental property of mathematics that is understood by starting with the natural numbers, the numbers that one can count on their hand. We begin with the number one or "unity" which is the 'mother' of all numbers. For example, a difference is necessary for there to be the number two. And the number three is necessary for proportions of numbers. Mathematically, self multiplication of any number, except the numbers 1 and 0, produces an exponential. If a number is divided equally by itself then a surprising thing happens called the "irrational numbers". In does not equal B. An important logical fact is that unity is indivisible, thus, the square root of 1 = ±1 and when one takes the square root number greater then or less than one then a class of numbers called "irrational numbers" are produced as defined by an infinite repetition of numbers to the right of the decimal point, e.g., square root of 2 = 1.4142.... In contrast to the rational number square roots, like the numbers 4, 9, 16, 25, 49, 64, etc that are unity by a module of the natural numbers (in mathematics, a module is a generalization of the notion of vector space in which the field of scalars is replaced by a ring). Greek mathematicians only used the positive natural numbers in their huge engineering and architectural accomplishments, because irrational numbers that extend to infinity were too troublesome and impractical to bother with. It was not until after the death of Christ that zero and negative numbers were used. It was not until the 1700s that complex numbers were conceptualized to include the entire number system, from counting numbers, natural numbers, rational numbers, irrational numbers and transcendental numbers in which negative numbers and the square root of negative numbers is calculated by rotating the number line. The Greeks conceived of numbers by mapping the positive numbers on a straight line like a ruler extended to infinity. The conceptual break through in the creation of complex numbers was to rotate the number line radially around zero in 90 degree steps by multiplying by the square root of -1. Thereby all numbers are represented on a two dimensional plane by two parts, one called the real part and the other called the imaginary part. This means that self-referential multiplication and division of numbers not only get larger or smaller, but also rotate to create circles and spirals and many curved shapes. Descarte was largely responsible for providing the visual representation of complex numbers and algebra in general. Vision is an important part of human consciousness and Descarte's marriage of mathematics and geometry unleashed the study of human visual brain networks, resulting in an explosion of mathematical thought that continues today. Suddenly, a relatively small number of complex numbers began to re-occur in human commerce and science; for example, the number 'e' = 2.71828 . . . in interest rates, missile trajectories, electricity, magnetism and calculus.[1]

---

[1] The mysterious number 'e' is the only number in the universe that is its own derivative, that is, the only number that is itself pure change. This remarkable number was first discovered in Egyptian times when computing interest rates of loans in commerce. The reason that 'e' is its own 1st derivative is because it is the series limit when the number one plus one divided by a number that is self-multiplied to infinity or the limit of $(1 + 1/n)^n$ as n approaches infinity. Unity is not divisible and 'e' is the number that is produced when one attempts to divide the indivisible using infinite regression of the division of unity by approximation to a limit. Interestingly, the limit of dividing unity by itself is the first derivative which itself is defined as the rate of change. That is, the failure of the process to divide unity yields only the process itself which is a basic and fundamental rate of change, i.e., 'e'. The number 'e' was used by Euler in the 1700s to create a complete closure of complex numbers that led to the development of a branch of mathematics called "Analytic Numbers" and "Conformal Maps". This branch of mathematics continues today in the Descarte vision of combining geometry with complex numbers in mappings of surfaces and spheres and higher dimensional spaces of the human brain and this branch of neuroscience is referred to as computational neuroanatomy as described by Schwartz, 1980.

1.6 WHY IS THE GOLDEN PROPORTION BEAUTIFUL?

The golden proportion is unique because it is the only number in the universe that is defined on a straight line segment by the proportion ab/ac : ac/cb, where c is a cut or point on the line ab which is the irrational number 0.628........ Thus, c is the golden section of AB (figure 1). This is the only section of a line segment in which the square of the larger segment produces the smaller segment. That is, i.e. a2 = b or conversely, the square root of the smaller segment produces the larger segment, i.e. $a = \sqrt{b}$. In this regard, the golden section utilizes the symmetrical operation of "self-multiplication" to create least effort growth in living things, such as the sunflower or snail shell, etc. As mentioned previously figure 1 illustrates the simplicity of the golden ratio and its relationship to the logarithmic spiral and the Fibonacci series.

Numerous psychological experiments have been published on the subjective preference of individuals toward one of the most pre-eminent of Pythagorus's numbers: "the golden proportion". The Golden Proportion is often referred to as the "Divine Ratio". As explained earlier, it was discovered by Pythagoras as a "cut" or section of a line segment in which the ratio of the shorter segment to the whole is a .6218... or the ratio of approximately 1/3, 3/5, 5/8.

This ratio had a profound influence on Greek architecture and on mathematical discoveries both in of ancient time and modern times. For example, the Feigenberg number in nonlinear dynamics approximates the Fibonanci series, and the golden ratio is an expression of mathematical beauty as evidenced by the large number of books written about the Golden Proportion over the last 100 years. The earliest scientific studies of the aesthetic value of the golden proportion were conducted by Fechner in 1876 (Ortlieb et al, 2020; Phillips et al, 2010). Fechner made thousands of ratio measurements of commonly seen rectangles, e.g. playing cards, windows, writing paper, book covers, desk tops, etc., and found the average rectangle approximated the Phythagoran number "Phi" or the "Golden Proprtion". Further, Fechner tested aesthetic preference by presenting various rectangular shapes and asking subjects to rank order or rate the shapes by their "aesthetic appeal".

More recent studies have generally replicated Fechner's results and add further support to the aesthetic value of the golden proportion (Benjafield, 1976; Svensson, 1977; Benjafield et al, 1980; Benjafield and Adams-Webber, 1976; Benjafield and Green, 1978). Svensson (1977) reported that psychology and art students, when instructed to divide a line at the point where the resulting line segments formed the most pleasing ratio, produced a ratio close to the Golden Ratio. Experiments by Benjafield et al (1980) showed that subjects draw the proportions of 1/2 and the golden section with less error than either 2/3 or 3/4.

In addition, the Golden Section, was drawn within, more frequently than other proportions. Also, an elaboration of the role of the Golden Proportion in interpersonal relations was recently presented by Benjafield and Adams-Webber (1976) and Benjafield and Green (1978). Green (1995) also published an extensive review of this scientific literature and concluded that there are real psychological preferences associated with the golden proportion when studies are conducted using careful methodologies.

The aesthetic appeal of the "Golden Proportion" is not limited to the visual arts. It is seen in music, architecture, mathematics as well as art. An enormous literature exists concerning the role of the "Golden Proportion" in many areas of what must be considered as a uniquely human feeling, i.e., the "Aesthetic Feeling". For example, the role of the Golden Proportion in music is characterized as a build-up of tension, with a release of tension or resolution approximately 3/5 or 5/8 of the movement with variations of this temporal proportion within different themes and chordal arrangements.

The role of aesthetics in architecture includes the Greek Parthanon and the Egyptian Pyramids in which the Golden Proportion formed the mathematical foundation of the design of these monuments (Ghyka, 1977). The role of the Golden Proportion in mathematics is seen in mankind's fascination and use of the "transcendental numbers" and "non-linear dynamics" both of which fundamentally involve the golden proportion.

Simplicity is at the heart of much of creativity and art. This gives rise to the "aesthetic moment" or the "aesthetic feeling" that we feel when we see a sunset, a beautiful flower or a spiral gallaxy. What is the aesthetic commonality between a sunset, a flower and a water spout? The answer is "simplicity". Simplicity is a link between the physiology of the brain to the laws of physics, a type of atomic mapping and atomic resonance.

As described previously the hypothesis is proposed that the basis of human aesthetic preference toward the golden proportion is a match between minimal energy forms in the external world with minimal energy forms inside the brain. The Golden Proportion is one of the most elegant and prevalent minimal energy forms and serves as a good example of how simplicity, harmony and proportion give rise to aesthetic feeling. The hypothesized minimal energy resonance is specific and mathematically related to non-linear dynamics and chaos in both physics and biology. One of the "kernals" of the non-linear dynamic of both the brain and the external world is the "Golden Proportion".

The mathematical form of the golden proportion gives rise to biological "Multiplication by Addition" which confers "economic advantage" to growth and change. The Golden Proportion is the general rule of "the recurrence of the same proportions in the elements of a whole". Based on these and other principals: **"Aesthetic Feeling Reflects the Match Between the Minimal Energy Forms of the Neurophysiological mappings of the Brain and the Minimal Energy Forms of the External World".**

Specifically, a particular and general feeling of beauty arises in consciousness due to a physiological "least effort" match of the golden proportion anatomy of the brain to the golden proportion form outside of the brain, because of the unifying principle of "least action". This hypothesis is presented in the physiological recognition of brain chemistry related to the "reward" chemicals of dopamine, noradrenaline, catechol amines, etc. which are often released in response to aesthetically pleasing events and that feelings of aesthetic pleasure can surely be released by totally independent and unrelated mathematical and physical forms.

It is not the purpose of this book to explore the neurophysiological bases for the general feelings of "pleasure" and "reward" because these subjects are covered in numerous texts and reviews. Rather, the purpose of this paper is to focus on only ONE very precise "Number" (the Golden Proportion) in the Universe and ask **"Why do We Experience Pleasure from this Number"?** It is assumed that pleasure chemicals (such as dopamine, noradrenaline, acetylycholine, seratonin, etc.) are also involved in the generation of aesthetic feelings in response to the perception of the "Golden Proportion".

In order to explore the relationship between the "Golden Proportion", minimal energy forms, and the reward brain chemistry it is necessary to briefly review certain historical mathematical concepts that the Egyptian and Greek mathematicians explored.

1.7 THE LOGRITHMIC SPIRAL

The logrhythmic spiral is one of the most common minimal energy forms seen in spiral galaxies, hurricanes, tornadoes, stirring a straw thru a liquid, etc. The human brain also includes the embryological development of a fluid flow log spiral in the mapping of the retina to the visual cortex, the mapping of the skin surface to the sensory cortex, and the cochlea of the inner ear to the auditory cortex. All three primary sensory systems share a common mapping format called a "Conformal Map" with the golden proportion as a fundamental basis function. For example, the retina is mathematically the same as a disk (W) with a logarithmic spiral mapping to the visual cortex (Z) where $W = \ln Z$ is called a "Conformal Map". Here all values between zero and infinity are a logarithmic spiral mapping (Schwartz, 1977a; 1977b; Tootle et al, 1982). A similar log spiral mapping occurs with sound and touch (Schwartz, 1977a; 1980). The driving forces of the aesthetic compass of our lives is a spatial match of the structure of the brain such that the golden proportion gives rise to a subconscious and sudden aesthetic feeling as an instantaneous "minimal energy" match between the laws of the universe and the neuroanatomy mappings of the human brain.

The "aesthetic" match of the golden proportion form of external shapes and sounds and time patterns is a type of computational neuroanatomy including multiplication by addition as represented by convergent synaptic activation in dopaminergic neurons and in excitatory loops that produce long-term potentiation (LTP) (Hebb, 1940). The primary aesthetic feeling is a least effort "reward" involving dopamine in the medial frontal lobes that is associated with emotions such as "joy", "pleasure", "surprise", "delight", "respect", "awe", "reverence", etc. The "aesthetic compass" is a special and enlarged attribute of the human primate. It is this directive force of the "aesthetic feeling", matching with the golden proportion that gives rise to much of our subconscious goal directed behavior. Humans are naturally attracted toward "beauty". Our perception of beauty often results in shifts in goals and perspective from which creative endeavors and positive decisions arise.

2

CHAPTER

NEUROAESTHETICS

Beauty of whatever kind, in its supreme development, invariably excites the sensitive soul to tears.

Edger Allan Poe

I find beauty in the continual shaping of chaos which clearly embodies the primordial power of nature's performance.

Iris Van Herpen

Art is the imposing of a pattern on experience, and our aesthetic enjoyment is recognition of the pattern.

Alfred North W hitehead

As a faculty member at NYU school of medicine I had the good fortune of meeting Eric Schwartz and George Chaiken who were trained in mathematics and physics. The three of us spent time examining sensory maps that remarkably obeyed the laws of fluid flow and topology, such as conformal maps (i.e., a mathematical function that preserves angle). In conformal map mathematics, a logarithmic spiral is like a snail shell, the human cochlea, a tornado, or a sunflower, etc. that all preserve angle as they radially grow in length (for details see chapter on Computational Neuroanatomy). For example, the mapping of the retina to the visual cortex is a logarithmic spiral conformal map; the cochlea is a logarithmic spiral; and the mapping of the skin surface to the somatosensory cortex is a logarithmic spiral. Thus, there is a common format across sensory-cortical maps that, because of the mathematical form of the logarithm and the mathematics of the "Golden Proportion", give rise to computational neuroanatomy by virtue of the mapping itself (Schwartz, 1977a; 1977b).

The log mapping of the retina to the cortex, for example, provides size and rotational invariance that are intrinsic to conformal mapping. The shape of a computer (laptop or a desktop) is irrelevant but the shape of fiber connections and mappings and re-mappings in the brain guided by DNA/RNA fluid dynamic unfolding of minimal energy forms during embryogenesis is a fundamental aspect of brain function. The finding of conformal maps that involve the Golden Proportion suggested a linkage between aesthetic feeling and the form of the sensory-cortical maps themselves. This is significant in brain science because aesthetics is such an important part of the human experience and representing a feeling of appreciation of beauty that comes effortlessly and immediately upon contact with certain objects, forms and sounds (Chatterjee et al, 2016). The emphasis here is on the concepts of "effortless"and "immediate". It appears evident that the immediate perception of an object of beauty involves a certain degree of matching between the form of the object (i.e. its proportional properties) and the form of the sensory organization of the brain. Limbic and reticular interactions contribute to the feeling of beauty, the aesthetic components are cortically analyzed and finally a "figure of merit" is assigned by the limbic system (e.g., nu. accumbens, amygdala, nu. basalis, etc.).

The Golden Proportion or the irrational number 1.618... is in the shape of a snail shell or sunflower or a rose or a water fall that also elicits a limbic-cortical aesthetic figure of merit because it is hypothesized that golden proportions constitute that class of objects with the least effort transduction by the primary sensory system itself (Schwartz, 1977a' 1977b: 1980). The ratio of frequencies of the human Electroencephalogram (EEG) also exhibit a Golden Proportion; in fact, this ratio appears to be critical in phase reset dynamics of the human EEG. As discussed in the chapter III on EEG and Aesthetics Pletzer (2010) argued that the Golden Section is the most irrational of all irrational numbers.

For example, if one does a Fourier transform of the digits to the sequence of digits to the right of the decimal point of any irrational number, (like the square root of 2 or square root of 7) then one finds repeating segments or sequences of numbers giving rise to peaks in the spectrogram. Experiments by Pletzer et al (2010) demonstrated that when phase shifts in the EEG are near to the Golden Ratio of EEG frequencies, then the probability of phase lock is at a minimum. The Pletzer et al (2010) study further supports the conclusion that complexity and effortless dynamics are minimal energy forms intrinsic to the cross-frequency coupling of EEG rhythms, which are important for aesthetic feeling at a given moment of time.

The relationship between aesthetic feelings and "complexity" is often defined in terms of "perceptual effort" (Nadal M. and Chatterjee, 2019). This linkage between the complexity of an object and perceptual effort is paralleled by a similar linkage between mathematics and the simplicity of the physical laws of the universe. For example, the complex logarithm function, which is characteristic of the global and local structure of the sensory mappings of the brain, may also be used to describe the pattern of electric or magnetic fields, the velocity flow of a fluid, or the distribution of a diffusing chemical reactant. The basic developmental reason for this commonality is that structures of these sort require minimal encoding.

That is, they represent the most parsimonious and economical methods of controlling dynamic flow, and in the case of living matter, the growth of form. The unifying and simple nature of these observations indicates that their commonality is ultimately an expression of the "Variational Principle" or the principle of "Least Effort" in physics. This principle, as reflected in the calculus of variations, is a description of the processes by which nature finds the path of least resistance, which is the solution of greatest elegance, simplicity and parsimony in the resolution of conflicting forces in nature (a sunset, a flower, a tornado or hurricane).

The mechanisms of aesthetics and, perhaps more generally, perception, may directly involve a similar linkage. In this case, simplicity and perceptual effort are the least effort expression for the growth of living forms (the Golden Proportion), and these are matched by the least effort expression for the evolution of the physical universe. In other words, a fundamental aspect of the human aesthetic feeling involves a match between the organizational laws of the atoms of the brain and the organizational laws of the atoms of the environment or space external to the brain. The mathematical variational principle of Euler-Lagrange and Hamilton is a universal expression that applies to living matter and human consciousness.

In the case of human consciousness, the negative entropy of the synchronous discharge of millions of neurons time linked to the present and matched and mismatched to memory and expectations of the future involve 80 to 300 millisecond intervals of time. These integrals are sequential minimal energy states. Another factor is that a fundamental neural process is phase resetting and phase realignment of billions of neurons that occurs with no net expenditure of energy. Thus the high speed recruitment of large numbers of neurons is itself a minimal energy form.

2.1 GOLDEN PROPORTION AND SENSORY MAPPINGS OF THE BRAIN

"The issue of the cortical movie screen, popular at first, discredited later... and defended once again, is still not resolved. The presence of these topographically organized projection areas can hardly be mere accident, of course. Besides the retina and the body surface, the receptor sheet of the cochlea also finds a representation of sorts in several regions of the brain. Whatkind of significance can we attached to them?"

(Somjen, 1972)

It is a well established fact that the neurons arising from the sensory surfaces of the body (e.g. the retina or the skin or the cochlea of the ear) project in an orderly fashion to their respective primary receptive areas in the cortex. This orderly projection of the peripheral sensory surfaces to the cortex is referred to as the topographic mapping of the brain. In the early 1900's these topographic maps were considered to be of physiological relevance in the processes of perception and sensation.

However, until approximately 1977 this view had fallen into disfavor, largely because until 1977 there had not been any definitive studies demonstrating that retention of the spatial form of a stimulus has physiological value. Simply put, the remapping of the peripheral representation of the sensory stimulus seemed to offer no special advantage to the organism.

Recently, research has resurrected this issue by suggesting a powerful computational role of the anatomical mappings themselves (Schwartz, 1977a; 1977b; 1980; 1984; Werner, 1970; Wieman and Chaiken, 1977; Cavanagh, 1978; Tootel et al, 1982). In particular, Schwartz (1977a; 1977b; 1980) has advanced a mathematical model of the visual sensory mappings which may provide considerable computational advantage to visual perception in general. This mathematical model, which is based on strong anatomical evidence, is a logarithmic conformal map that uses the formula of the logarithmic spiral as its basic function (i.e., $r = \alpha e^{k\theta}$).

In the sections to follow, evidence for the logarithmic mapping (i.e., $r = \alpha e^{k\theta}$) of the body periphery to the central nervous system will be presented for three of the major sense modalities (i.e. vision, hearing and touch). In the present paper the visual system will be emphasized; however, it will be argued that all three sense modalities involve a log spiral mapping and that this mapping is of fundamental importance to perception and to the brain mechanisms of aesthetics. Further, it will be argued that an "aesthetic metric" can in theory be developed which relates the magnitude of aesthetic feeling to the degree of match between the geometric properties of the physical world (using the Golden Proportion as the basis function) and the geometric properties of the sensory systems of the brain (which also use the golden proportion as a basis function).

2.2 VISUAL RETINO-CORTICAL MAPPING

A brief summary will be given here of the mathematical formalities presented by Schwartz (1977a, 1977b, 1980). An important concept in the retinotopic map is the concept of a "magnification factor" introduced initially by Daniel and Whitteridge (1962). The cortical magnification factor is the distance on the visual cortex which corresponds to the distance moved by a spot of light across the surface of the retina. This quantity can be expressed by the equation: $m = k/r$, where k is a constant, r represents eccentricity in degrees from the fovea, and m is the magnification in millimeters/ degree. The cortical magnification is a differential quantity. By this is meant that a small change in cortical position is related to a small change in the visual field position.

Polyak (1941) was the first to suggest, on the basis of the anatomy of the visual cortex, that a mathematical projection of the retina to the cortex must exist. Talbot and Marshall (1941) confirmed this hypothesis with physiological data. However, it was not until the extensive investigation of Daniel and Whitteridge (1961) that a precise, quantitative source of data was provided. Daniel and Whitteridge found that the cortical magnification factor is the same along all radii, regardless of the angular coordinate, that is, it is the same in all directions. Schwartz (1977a) investigated the properties of the Talbot and Marshal (1941) and Daniel and Whitteridge (1961) studies in great detail. He concluded that a simple and accurate description of the retinotopic visual map was provided by an analytic function whose derivative is radially symmetric and proportional to 1/r. The only such function is the complex logarithm illustrated in figure 5: or W = ln Z.

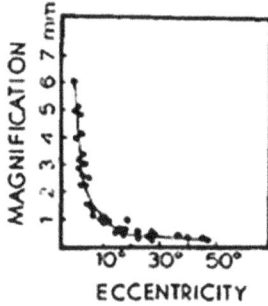

A

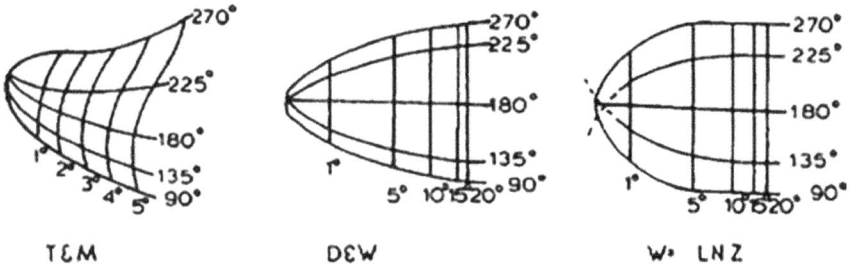

B

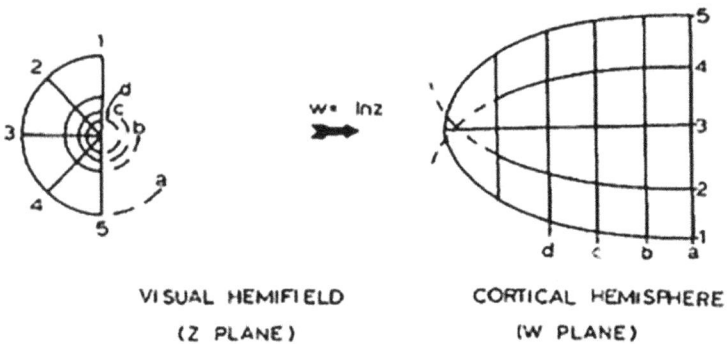

C

Fig. 5-A. The cortical magnification data of Daniel and Whitteridge. Through the points is drawn the best fit to the data for a power law. B. The measured and predicted mapping of visual landmarks in striate cortex. The upper (90o) and the lower (270o) vertical meridians, the horizontal half meridian (180o), the octants (135o and 225o) and the circles of constant eccentricity are drawn as measured by Talbot and Marshall, and Daniel and Whitteridge. The data of Talbot and Marshall, on the left, does not show the correct (logarithmic) spacing between the lines of constant eccentricity; because their experiment was the pioneering measurement of this data. The data of Daniel and Whitteridge is much more accurate, and is shown in the center. This is a projection, onto a horizontal plane, of a three dimensional model; the meridians and octants are equally spaced, as they are in the theoretical prediction of these mappings under the logarithmic conformal mapping. The theoretical prediction, on the right, actually represents a vertical meridian that is infinitesimally displaced from the origin; otherwise the curved part of the contour would actually be a right angle. The horizontal meridian is an average of a line infinitesimally above and below the precise horizontal meridian. With these qualitative reservations, there is a great similarity between the data and the theoretical prediction of the data under the logarithmic mapping. C. The global retinotopic mapping under the logarithm function. Concentric circles (exponentially spaced) and radial lines are mapped onto the equidistant cartesian grid on the cortex. Note the density (derivative) of the exponentially spaced lines gives a linear dependence on the eccentricity; this is observed as a linear scaling of the receptive field size in the visual plane, with a constant (hypercolumn) size in the cortex (from Schwartz, 1977a).

Analytic functions or conformal mappings are defined in several equivalent ways (Ahlfors, 1966) of which Schwartz emphasizes two. The first is that the magnification factor in the vicinity of a point on the retina is independent of direction. Daniel and Whitteridge (1961) have shown that this is true, based on their findings that the value of the magnification factor remains constant, no matter which direction the spot of light is moved on the retina. A second requirement for an analytic function is that the direction and magnitude of local angles are preserved (Ahlfors, 1966). Thus, in a conformal map of the retino-cortical mapping, a set of lines resulting in a right angle on the retina will also result in lines which intersect at right angles in the cortex. A particular angle on the retina is retained, through the logarithmic conformal mapping, so that it is represented locally in the cortex even though on the global scale there is a large distortion. This fact allowed Schwartz (1977a) to argue that the computational operation of size invariance and rotational invariance is an intrinsic property of the retinotopic projection system.

The logarithmic conformal function involves essentially a mapping of a disk onto a rectangle as shown in figure 6. As can be seen, radial lines in the retina

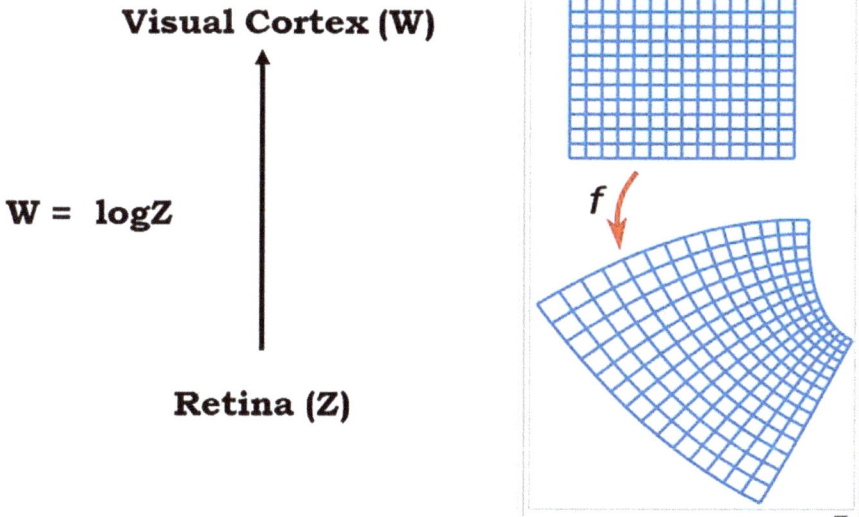

Visual Cortex (W)

$W = \log Z$

Retina (Z)

f

Fig. 6 - A rectangular grid (top) and its image under a conformal map f (bottom). It is seen that f maps pairs of lines intersecting at 90⁰ to pairs of curves still intersecting at 90⁰ . This figure also illustrates the conformal map of a disk (retina) to a rectangle (visual cortex) where the angles are preserved resulting in warping of space. The space-time warping of light by planets and galaxies is mathematically also described by conformal maps. From https://en.wikipedia.org/ wiki/Conformal_map

are represented in the cortex as parallel horizontal lines, while concentric circles in the retina are represented in the cortex as parallel vertical lines. As mentioned previously the formula for the logarithmic spiral $r = Ae^{k\theta}$ generates straight lines and circles at the extreme limits of $r = Ae^{k\theta}$ (i.e. all values between 0 and infinity.

Figure 7 summarizes the geometric properties of the conformal map in different species. Figure 8 shows a series of graphical fits of the complex logarithmic function of the mapping of the retina to the cortex for several different species of monkeys and for the upper visual field of the cat. This single and simple analytic function provides a good fit to the experimental data from various species. Of course, it is assumed that human data, once acquired, will be similar to that of the non-human primates.

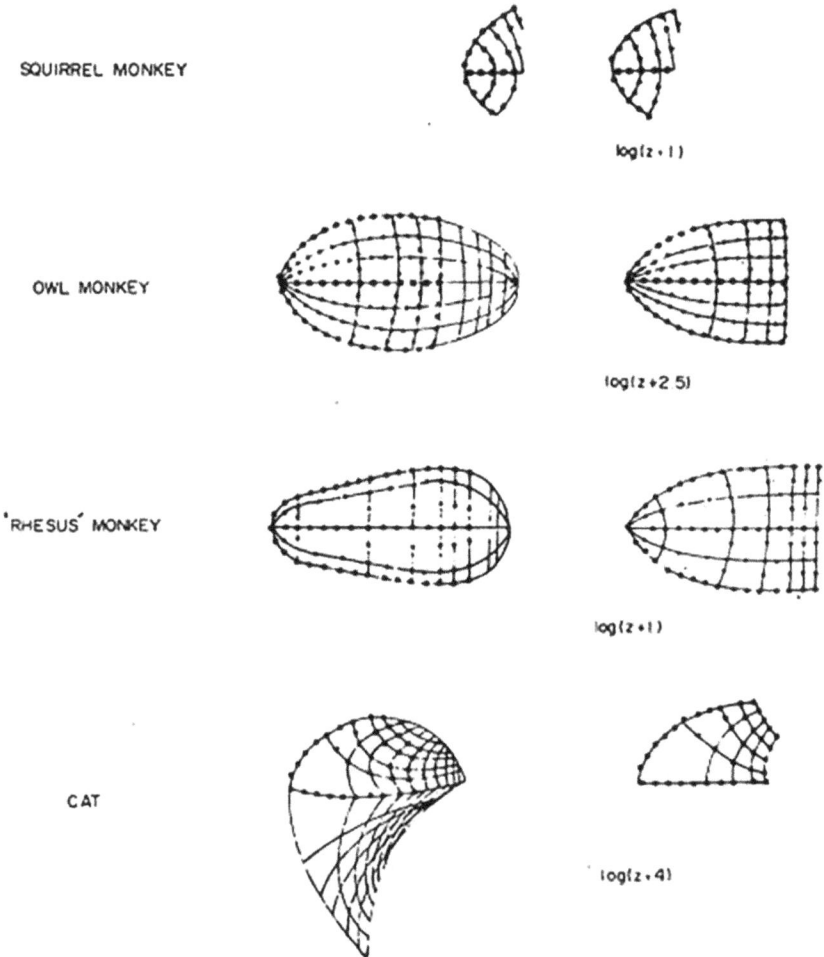

SQUIRREL MONKEY

$log(z+1)$

OWL MONKEY

$log(z+2.5)$

'RHESUS' MONKEY

$log(z+1)$

CAT

$log(z+4)$

Fig. 7 A series of computer generated complex logarithm conformal mappings which provide the best (visual) fit to published retinotopic (striate cortex) mappings in a number of primate species, as well as the cat. The experimental mappings are shown in the column labeled "experiment", and the complex logarithmic mappings in the column labeled "theory". Only the central 20-40⁰ are presented in the theoretical maps, and the corresponding areas of the experimental maps have been emphasized with the following graphic symbols: circles mark the projection of the vertical meridian, and squares mark the projection of the horizontal meridian. In the cat, it appears that the complex logarithmic approximation is quite good for the upper visual field, as shown, but fails to represent the lower visual field. For the "rhesus" monkey map (Daniel and Whitteridge, 1961). the experimental map was drawn as an "orthogonal projection", rather than as a "flat map", which is the case for the other data. This accounts for the lack of curvature of the circles of constant eccentricity (compare with the owl monkey (Allman and Kaas, 1971) map, or the theoretical maps. Also, in this work, a mixture of different primate species, in addition. to rhesus monkeys, were used, although only one map was published; this fact is acknowledged by the use of the quotation marks (i.e. "rhesus"). In summary, this figure demonstrates that it is possible to provide a simple analytic approximation to the global retinotopic mappings of a number of different species, in terms of the general form of the complex logarithm of a linear function of visual field coordinates. In the case of the squirrel monkey (Cowey, 1964) only

the central 4⁰ of cortical map have been published
as a flat map; however, the magnification factor of
squirrel monkey and rhesus monkey have the same
functional form. From Schwartz, 1980).

Table I explains the geometry of the logarithmic conformal mapping.

Table I

Log z	=	W
1. Concentric circles **(exponentially spaced)**		**Vertical lines** **(equally spaced)**
2. Radial lines **(equal angular spacing)**		**Horizontal lines** **(equally spaced)**
3. Logarithmic Spirals **(G = Aᵏθ)**		**Inclined straight lines** **Slope = 1/k; intercept = -log A/k**

Table I-The three geometric patterns on the left are
the level lines, or streamlines, of the logarithmic
conformal mapping. The W mappings of log z are
described on the right.

Figure 8 illustrates how concentric circles in the Z plane (retina) map to vertical lines in the W plane (Cortex), while radial lines in the Z plane (retina) map to horizontal lines in the W plane (Cortex); and finally a logarithmic spiral in the retina is represented in the cortex as an oblique straight line.

Z Plane   W Plane

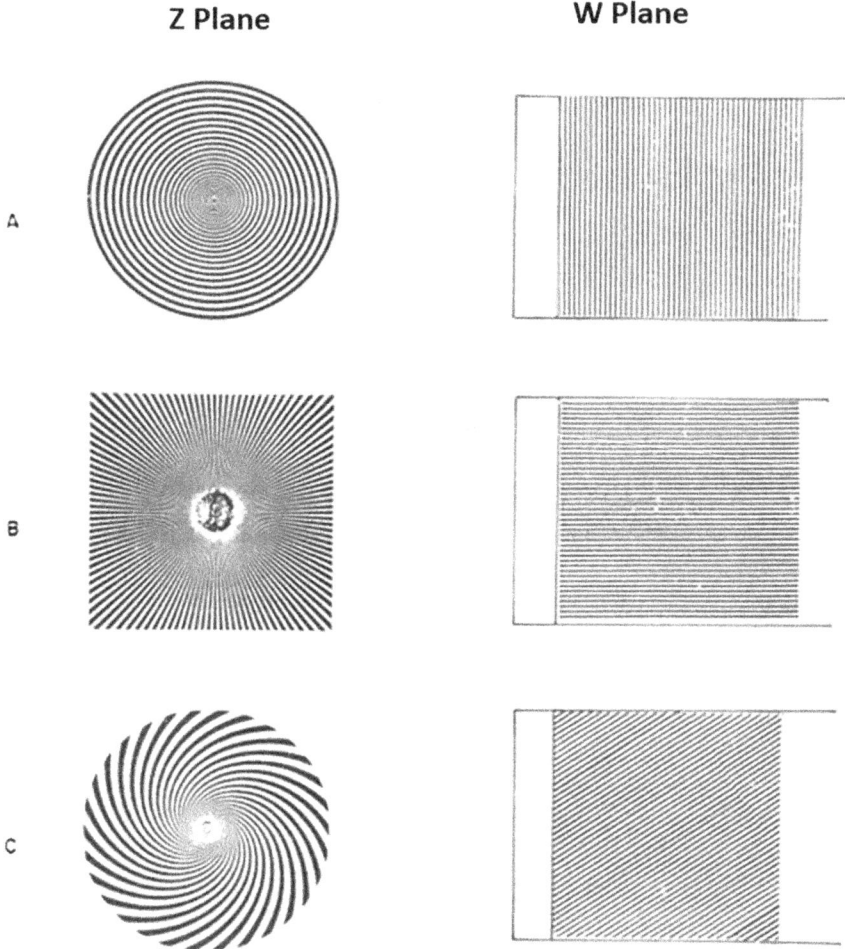

A

B

C

Fig. 8-The patterns on the left (A, B, C) are examples of MacKay complimentary stimuli (A) and (IS) are complementary while (C) is complementary to a log spiral of opposite "handedness". The images of these stimuli, under the map w = log(r) are shown on the right. The central circle (singularity of the logarithm function) is omitted in each case, and in fact, is omitted from MacKay's figures due to limitations of the printing process. This figure indicates that parallel rectilinear grids [or equivalently, sinusoidal gratings) are associated with MacKay grids, via the logarithmic mapping The grid stimuli on the left are typical eigen patterns of the Mellin-Fourier transform. It is clear that MacKay complementary afterimages have a dose relationship in the cortical plane. From Schwartz, 1980.

Figure 9 is a graphic simulation of the geometrical properties of the complex logarithmic mapping. in terms of a series of deformations of an imaginary "plastic" material.

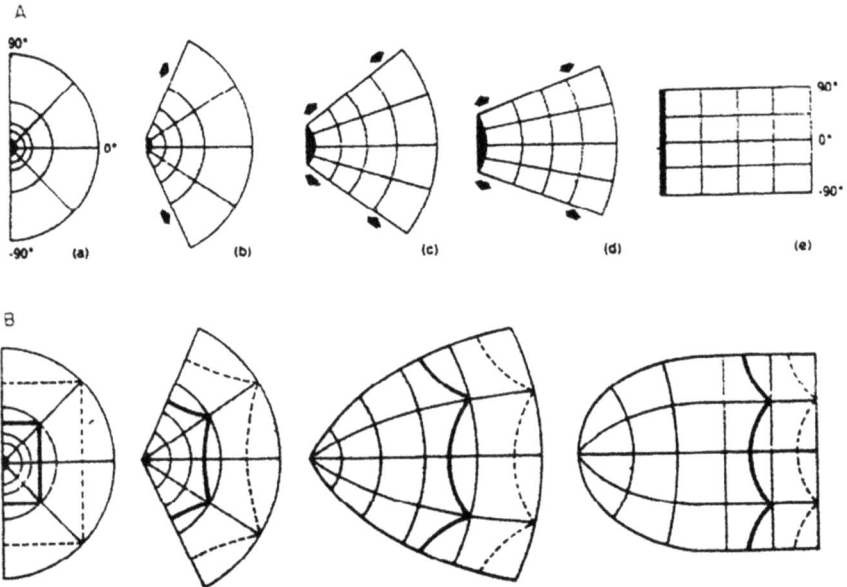

Fig. 9 Graphic simulation of the geometrical properties of the complex logarithmic mapping. in terms of a series of deformations of an imaginary "plastic" material. On the top left, a logarithmic "radex" is drawn. In figures (b)-(e), this radex, which may be identified with the retina, or the visual field,

58

is smoothly deformed so that its final state, (e), represents the complex logarithmic mapping of the radex. The exponentially spaced concentric circles of (2) have been mapped into parallel. equi-spaced vertical lines; the rays of (2) have been mapped into parallel, equi-spaced horizontal lines. The central black circle of (a) has been stretched into the black band of e). This black circle represents the singularity of the logarithm function. In the lower part of the figure, the singularity is removed by using as mapping function log (1 + Z). This mapping is quite similar to the logarithm, except at Z = 0, where it is finite. Also shown in the figure are the deformation of a large and small square. under this mapping. It can be seen that the change of shape induced by the mapping is exactly such as to cause the final images to be identical, in size and shape. A similar property holds for rotation. This is the basis of the pseudo-invariance properties of the complex logarithmic mapping discussed in he text. From Schwartz, 1980.

Figure 10 is an illustration of how a large and a small square, on the retina, would be transformed through the conformal mapping on the retina into two identical forms in the cortex. It can be seen in this figure that the shapes of the two squares remain the same although their size is different.

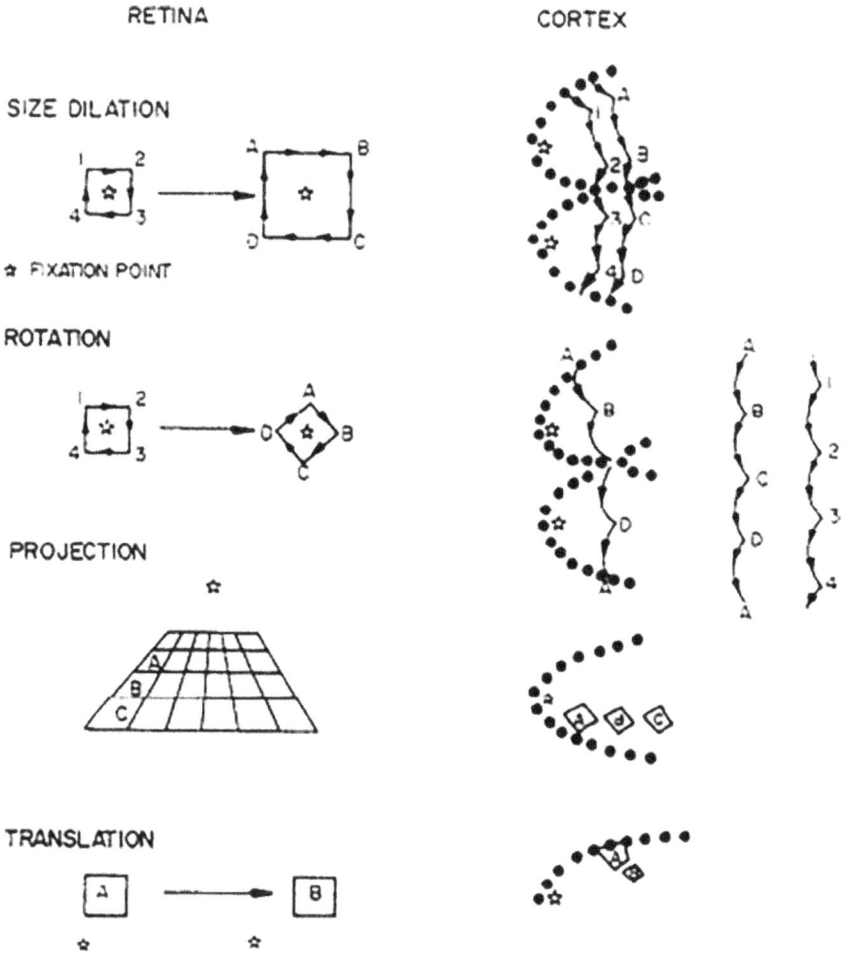

Fig. 10 Transformation properties of the complex logarithm function, under size, rotation, and translation. On the top left are shown two stimuli: a large and small square. The fixation point is represented by a star. On the right is shown the mapping under the function log (z + 1) of these two stimuli. The squares are assumed to subtend 20 and 40 of visual field. It is clear that the cortical images of these para-foveal stimuli, which differ in size

by 100%. are essentially similar in size and shape. (The left and right cortical images have been drawn together on the right.) A similar property is shown for rotation below. Size and rotation, in the complex logarithmic plane, are converted to shift, as described in the text. This property forms the basis for recent applications of complex logarithmic mapping in computer and optical pattern recognition. Also shown in the figure is the "projection" of a rectilinear grid, with the star representing fixation. The cortical mapping is suggested on the right, indicating that projection symmetry, which is in this case a special case of size symmetry, is also "normalized" by the complex logarithm. This suggests that the radial size variant flow of visual stimuli, during movement. may be converted to a rectilinear. size invariant flow at the cortex, for certain conditions of relative motion of the observer and the stimulus. Finally, at the bottom of the figure, translation is shown. Retinal images are distorted in both size and shape by the cortical map, although local angles are preserved. due to the conformal property of the mapping (From Ahlfors, 1966; Schwartz, 1977a).

However, their relative positions on the retina would be slightly different. Notice that the mapping of the larger square onto the cortex is displaced slightly to the right of the smaller square. This property of invariant shapes for different sizes has been suggested by Schwartz (1977a) to provide the basis for the perceptual phenomenon of "size invariance".

Additional experimental evidence in support of this particular mapping is provided by studies by Allman and Kaas (1974) for the secondary visual cortex of the owl monkey.

Their data is reproduced in figure 11. It is evident from this figure that the cortical image, under a straight line across the surface of visual cortical area II, is a spiral pattern of receptive fields, and that the mapping images this spiral mapping onto a straight line as per the logarithmic conformal mapping.

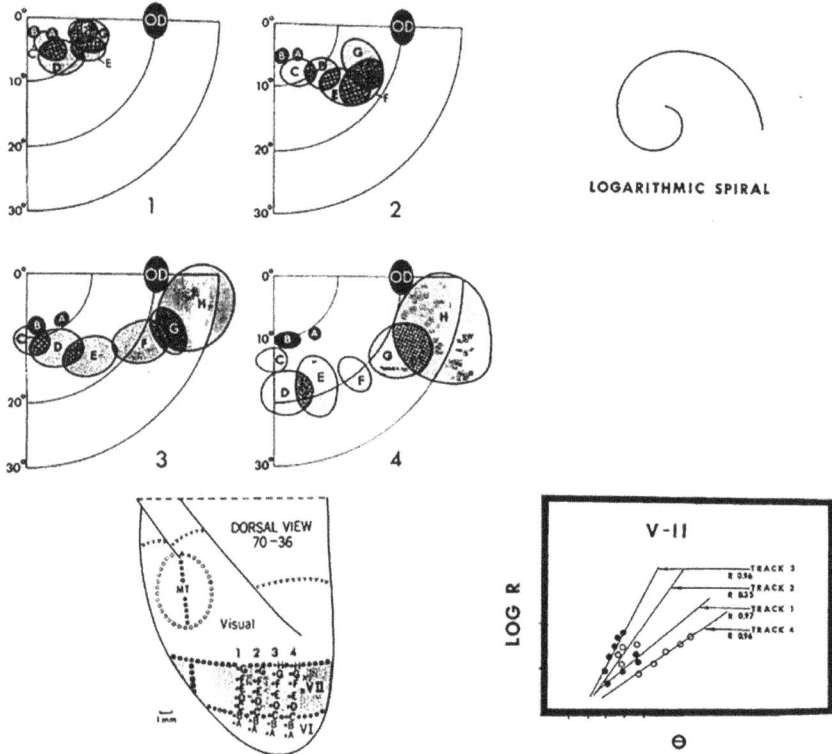

Figure 11-On the left is reprinted the data of Allman and Kaas showing their results for the measurement of receptive field size and position, corresponding to straight lines across the secondary visual area of the monkey. The perimeter charts labeled 1,2,3, and 4 correspond to the anatomical locations indicated in the lower part of the figure. On the right is an example of a logarithmic spiral. Below the spiral

is a semi-logarithmic plot of the radial position of the receptive field centers with respect to the corresponding angular positions. The hypothesis that these receptive field trajectories lie along logarithmic spirals is equivalent to the hypothesis that this semi-logarithmic plot should be linear. The coefficients of linear correlation to the best (least-squares) fit to a straight line are shown in the figure. The measurements were made directly from the figure of Allman and Kaas (1974).

Recently, Tootell et al (1982) directly tested the formalism of Schwartz (1977a) in a radioactive 2-deoxyglucose study of primate visual cortex mapping. In this study, monkeys were first injected with radioactive 2-deoxyglucose and then they stimulated with logarithmically spaced circles and equiangular rays. After approximately 20 minutes of visual fixation, the monkeys were sacrificed and the brain removed to determine the areas of maximum radioactivity in the visual cortex. The logarithmic conformal mapping model predicts that logarithmic spaced circles and equiangular rays on the retina project to an approximately rectangular pattern at the level of the visual cortex. Figure 12, shows the results of the Tootell et al findings, in which a remarkably good fit to this model was observed. The gross features of the logarithmic conformal map are dramatically evident in the Tootell et al (1982) paper (see fig. 12 -14).

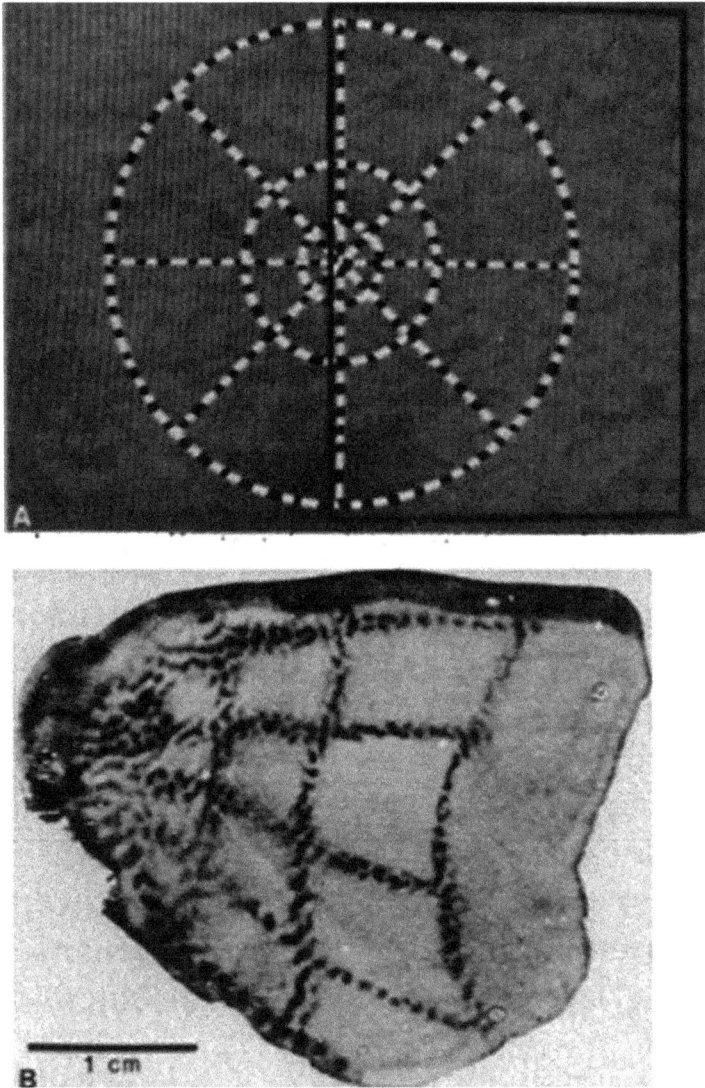

Fig.12-(A) One of the visual stimuli used. The solid black rectangle encloses that portion of the visual stimulus that stimulated the region of striate cortex shown in (B). (B)Pattern of brain activation. Produced by the visual stimulus shown in (A), as revealed by 2DG. This is an auto radiograph from a

single flat mounted tissue section (mostly from layers 4B and 4C). About half of the total surface area of the macaque striate cortex can be seen.

2.3 LOCAL GEOMETRIC STRUCTURE OF CORTICAL HYPERCOLUMNS

An important computational and aesthetic nesting property is that the global cortical log spiral map is recapitulated in the orientation columns of the visual cortex. A common feature of the golden proportion is that the retino-cortical mapping exhibits exquisite and elegant recursive properties which can be used to describe the local mapping of lines within a "hypercolumn". Schwartz (1977a; 1977b; 1980) showed that this property is demonstrated by the fact that a hypercolumn is composed of 12 to 18 orientation columns, each of which is tuned to a specific orientation of line in the visual field. The orientation columns exhibit the property of "sequence regularity" in which each orientation column is tuned to an angular range of 10 to 15 degrees, with successive columns representing a full 180 degrees of visual space. In other words, retinal radial lines that span a full 180 degrees in visual space are mapped to a rectangle of cortical slabs or columns. This mapping of "local" retina to local cortex is represented in figure 13. In figure 13 it can be seen that a line stimulus presented in equi-angular concatenated local columnar organization are embedded in the global organization of the cortical hypercolumns. Linear steps on the retina will map to equally spaced

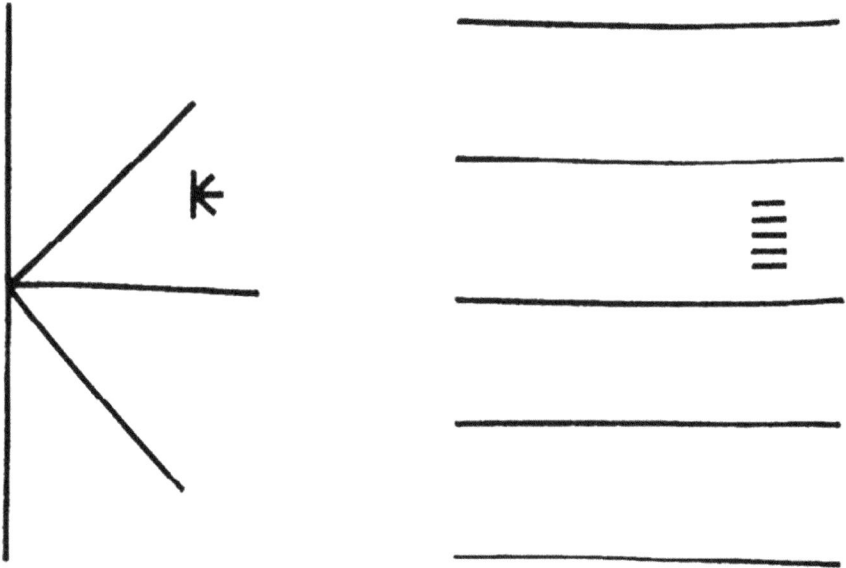

VISUAL FIELD CORTEX

Fig. 13-Local and global geometric structure of the cortex in terms of a simple graphic. This figure represents the fact that, on a global scale, equal angular rays are mapped to (approximately) parallel, qua1 spaced lines in the cortex (as in Figs 1 and 2). This geometric property is repeated on the local scale, since equal angular steps in the orientation of edges in the visual held map to parallel, equal spaced slabs in the cortex. This is simply a statement of the sequence regularity property of the hypercolumn model of Hubel and Wiesel(1974). The geometric property of mapping equi-angular rays to parallel slabs is the characteristic signature of the complex logarithmic mapping (Schwartz 1977a). Thus, the cortex may be thought of as a concatenated, complex logarithmic map. The geometry of the whole is repeated in the small. This image of the cortex as consisting of roughly 3000 complex logarithmic maps (hypercolumns), arranged in a global complex logarithmic pattern, suggests that the retino-cortical system might be described as a logarithmic compound eye. From Schwartz, 1977a.

columns in the cortex. This analysis indicates that there is a economy or parsimony of form that is present since the developmental rules which are responsible for shaping the global structure of the cortex are also sufficient to specify the local structure of the cortical map (Schwartz, 1977b).

Further, any functional advantages that are associated with the complex logarithmic mapping are available at both the local and global levels. Concatenated structure is frequently seen in biology, wherever an organism consists of a number of similar parts, and in which the development of the parts repeats more or less exactly the development of the whole organism (i.e. gnomic structure). In this context, the logarithmic spiral that describes the retinal-cortical map is a simple expression of a common biological growth law.

2.4 COMPUTATIONAL ADVANTAGES OF THE LOGARITHMIC MAPPING

The functional advantages of logarithmic spiral sensory cortical map are important to understand the biological relevance, simplicity and economy of the Golden proportion spiral in terms of visual perception. Among the potential advantages are: 1-size invariance, 2-rotational invariance, 3-information compression, 4-depth perception, 5-form perception, 6-visual illusions and 7-inter-sensory comparisons (Schwartz, 1980). All of these advantages of perception are a consequence of the intrinsic properties of the logarithmic conformal map. Each of these areas is discussed in detail by Schwartz (1980; 1984).

It is remarkable that a single and relatively simple model can be used to solve an assortment of visual perceptual problems. Several predictions which have arisen directly from this model have already been confirmed. For example, at the time of Schwartz's first paper there was no evidence that binocular-disparity tuned neurons existed (see Hubel and Wiesel, 1974). However, Poggio and Fischer (1977) confirmed that they did indeed exist and that they were attuned to about 0.05 degrees as, predicted, in advance, by Schwartz (1976). A second prediction of Schwartz (1977b) was that the ratio of length of a cortical hypercolumn to its width should be about 0.28. This very precise quantitative prediction has also been confirmed (Strykker et al, 1977). A third prediction concerns the hypothesis that near the center of the fovea ($z = 0$), the orientation column pattern is weakened, due to the singularity near $z = 0$. At the time that this model was proposed (Schwartz, 1977a) there was no evidence of orientation tuning weakening available. In fact, the data of Hubel and Wiesel (1962; 1974) suggested continuous bands of iso-orientation columns.

Later, however, Hubel and Livingstone (1981) have reported that near the center of individual ocular dominance columns, there is a noticeable deficit of orientation tuning. This observation is in good agreement with the logarithmic conformal model. Finally, in 1982 (Tootle et al, 1982) positron emission tomography (PET) studies of human visual cortex confirmed another of Schwartz's predictions, i.e., that circles in the retina map to straight lines in the visual cortex which is a log spiral conformal map.

2.5 SOMATOSENSORY CORTEX

Neurons from the cutaneous periphery synapse first in the spinal cord and the gracil or cuneate nuclei before ascending via the dorsal column-medial lemniscal pathway to the thalamus and the cortex. The cutaneous periphery is represented by a map-like representation of the body, as viewed from neurons in the post-central gyrus of the cerebral cortex (S-1). This is called the "somatotopic map". The mathematical conformal nature of the somatotopic map was revealed by Werner and Whitsel (1968) who measured the projection of straight lines of cells in the cortex (S-1) to the surface of the limbs. They found that "the receptive fields of the neurons progress, essentially, in bands around the limb, much as did the laces of a Roman soldier's footwear...the sum total of all receptive fields represented in any mediolateral traverse of the cortical map describes a continuous "spiral" (italics mine) path around the limb".

This observation, when coupled with the fact that the size of the cutaneous receptive fields increases linearly with distance from the distal point of the limb (Mountcastle, 1957) indicates that the somatotopic mapping takes straight lines in the cortex to logarithmic spirals in the cutaneous periphery. Werner and Whitsel (1968) state further that for rostro-caudal trajectories across the surface of S-1, "the sequence of receptive fields describe circular paths around the limb". As noted previously the above descriptions are that of a logarithmic conformal mapping for the somatotopic sensory system, with the map centered about the distal point of the limb.

In summary, the parallels between the visual and somatic maps are: 1-the receptive field size for the visual and somatic map scales linearly with distance from the distal point of the receptor surface; and 2-straight lines in the cortical representation correspond to receptive field trajectories that are concentric circles, logarithmic spirals, or radial lines. In addition, the motor representation of the cortex is itself a mirror-image of the somatotopic representation. As Schwartz (1977a) states, "Thus, the visual, somatotopic, and motor maps of the primary cortical representation may be described, at least approximately, by the same mathematical function: the complex logarithm" which is a log spiral like Galaxies, Tornadoes and snail shells.

2.6 AUDITORY CORTICAL MAPPING

One of the most aesthetically pleasing geometric structures in the body is the "cochlea" (located in the inner ear) which exhibits a logarithmic spiral structure. Von Bekesy (1953) developed a place theory of hearing in which temporal information (frequency) was represented in terms of the spatial distribution of frequencies in the basilar membrane of the cochlea. Subsequent biophysical analyses have shown that wave models, involving the fluid dynamics in the cochlea as well as the basilar membrane, combine to code frequency as a function of space. Studies by Honrubia and Ward (1968) show that in the cochlea, the point of maximum sensitivity, indicated by electrical measurements, exhibits a spatial displacement as a function of the logarithm of the frequency.

Microelectrode studies of the brain of the cat (Merzenich et al, 1975), squirrel (Merzenich et al, 1976), and monkey (Merzenich and Brugger, 1973) indicate that the cortical tonotopic map is essentially logarithmic. The study by Romani et al (1982) used neuromagnetic techniques to gain high resolution measurement of the dipole activity in the human cortex which was elicited by different frequencies of auditory stimuli and discovered that human tonotopic map to the cortex is logarithmic.

The functional value of such a logarithmic tonotopic map may be related to the fact that the just noticeable frequency difference within the bandwidth 500 to 5,000 HZ, is a fixed percentage of the frequency. In other words, an invariant is computed by the logarithmic transform such that the least noticeable decrement of the logarithm of the frequency is a constant, independent of frequency. Further, if we assume that the neurons of the auditory cortex have a uniform width and density, then the tonotopic logarithmic map implies that the same number of neurons in the cortex are dedicated to each octave change in frequency. This is another example of simplicity and minimal energy dynamics.

2.7 AESTHETICS AND THE PLEASURE NETWORK

Kawabata and Zeki (2004) used fMRI to evaluate whether there are brain areas that are specifically engaged when subjects view paintings that they perceive to be beautiful, regardless of the category of the painting (that is whether it is

a portrait, a landscape, a still life, or an abstract composition).
They concluded:

> "The results show that the perception of different
> categories of paintings are associated w ith distinct
> and specialized visual areas of the brain, that
> the orbito-frontal cortex is differentially engaged
> during the perception of beautiful and ugly stimuli,
> regardless of the category of painting, and that the
> perception of stimuli as beautiful or ugly mobilizes
> the motor cortex differentially."

Below are examples of neural activity resulting in changes
in the blood flow of the brain in different brain network regions
or regions of interest that are also known network hubs such
as the orbital frontal network for beautiful stimuli versus the
motor cortex for disgusting and ugly stimuli. One conclusion
is that ugly and disgusting images and experiences result in
activity in avoidance or repulsion motor actions. In contrast,
beautiful stimuli activated the ventral and orbital frontal lobes
with limbic connections but no motor activation or avoidance.
Salimpoor et al (2013) used functional magnetic resonance
imaging to investigate neural processes when music gains
reward value the first time it is heard. The degree of activity
in the mesolimbic striatal regions, especially the nucleus
accumbens, during music listening was the best predictor
of the amount listeners were willing to spend on previously
unheard music in an audition paradigm. Importantly, the
auditory cortices, amygdala, and ventromedial prefrontal
regions showed increased activity during listening conditions
requiring valuation, but did not predict reward value, which
was instead predicted by increasing functional connectivity
of these regions with the nucleus accumbens, as the reward

value increased. Figure 14 illustrates how this indicated that aesthetic rewards arise from the interaction between mesolimbic reward circuitry and cortical networks involved in perceptual analysis and valuation.

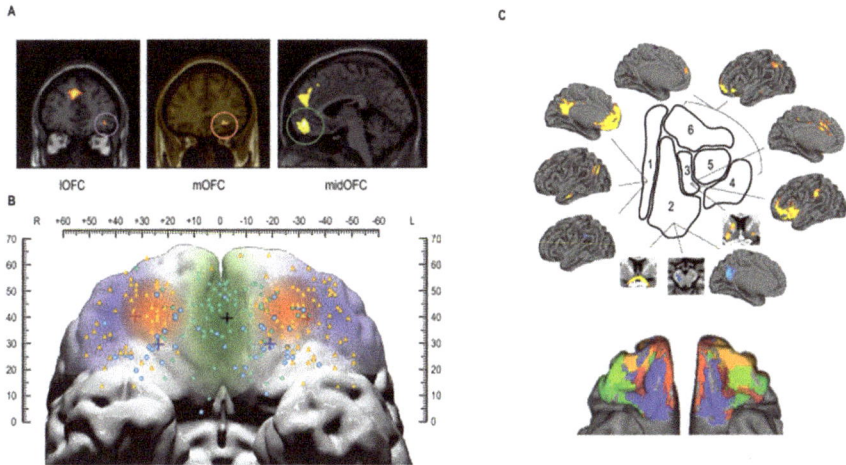

Fig. 14-Hedonic coding in the human orbitofrontal cortex (OFC) In humans, the orbitofrontal cortex is an important hub for pleasure coding, albeit heterogeneous, where different sub-regions are involved in different aspects of hedonic processing. A) Neuroimaging investigations have found differential activity to rewards depending on context in three subregions: the medial OFC (mOFC), mid-anterior OFC (midOFC) and lateral OFC (lOFC). B) A meta-analysis of neuroimaging studies showing task-related activity in the OFC demonstrated different functional roles for these three subregions. In particular, the midOFC appears to best code the subjective experience of pleasure such as food and sex (orange), while mOFC monitors the valence, learning and memory of reward values (green area and round

blue dots). However, unlike the midOFC, activity in the mOFC is not sensitive to reward devaluation and thus may not so faithfully track pleasure. In contrast, the lOFC region is active when punishers force a behavioural change (purple and orange triangles). Furthermore, the meta-analysis showed a posterior-axis of reward complexity such that more abstract rewards (such as money) will engage more anterior regions to more sensory rewards (such as taste). C) Further investigations into the role of the OFC on the spontaneous dynamics during rest found broadly similar sub-divisions in terms of functional connectivity (Kahnt et al., 2012) with an optimal hierarchical clustering of four to six OFC regions. This included medial (1), posterior central (2), central (3) and lateral (4– 6) clusters with the latter spanning an anterior-posterior gradient (bottom of Fig 3B), and connected to different cortical and subcortical regions (top of Figure 3B). Taken together, both the task-related and resting-state activity provides evidence for a significant role of the OFC in a common currency network. It is also compatible with a relatively simple model where primary sensory areas feed reinforcer identity to the OFC where it is combined to form multi-modal representations and assigned a reward value to help guide adaptive microinjections in NAc hotspot and coldspot. (red/ orange dots in hotspot = >200% increases in 'liking' reactions; blue dots in coldspot = 50% reductions in 'liking' reactions to sucrose). Panels show separate hedonic effects of mu opioid, delta opioid and kappa opioid stimulation via microinjections in NAc shell on sweetness 'liking' reactions. Bottom row shows effects of mu, delta or kappa agonist microinjections on establishment of a learned place preference (i.e., red/orange dots in hotspot) or place avoidance (blue dots). Surprisingly similar patterns of anterior hedonic hotspots and posterior suppressive coldspots are seen for all three major types of opioid receptor stimulation. From Castro and Berridge, 2014.

2.8 OPERATION OF CONSCIOUSNESS VS CONTENT

A current challenge of neuroscience is understanding the difference between the "Operation of Consciousness", such as sleep and wakefulness, versus the "Content of Consciousness" also referred to as "qualia", that is the subjective feelings and the personal and subjective world identified as the self. A recent study of the transition from propofrol unconsciousness to consciousness using an array of EEG electrodes placed directly on the frontal lobes referred to an an electrocortiogram (ECoG) demonstrated a special time scaling that approximates the Golden Proportion (Boussen et al, 2018). These authors used Time-Frequency analyses to compare the EEG propofol Unconscious (U) state during the transition to the EEG Conscious (C) state after the propofol was discontinued. Pre vs post propofol anesthetic induction shows a sharp decrease in gamma high frequency amplitude (> 20 Hz) and an increase in lower frequencies (e.g., delta 1 -4 Hz) as well as an increase in the coupling between delta/theta frequencies and gamma frequencies at the moment when consciousness is lost (Breshearsa et al, 2010). Awaking from the anesthetic is a reverse of the induction process where the state of being fully aware and conscious is associated with the reduced cross-frequency coupling of gamma

frequencies and theta frequencies across large-scale cortical networks (John, 2005; Lee et al, 2009). The default and attention networks are still operative during loss of consciousness which is believed to be necessary to maintain continuity of memory when consciousness is regained. However, reduced short-term memory capacity is prolong upon waking and most patients fail to remember instructions after they awake and this is why a parent, spouse, friend or guardian is asked to be present when post-operative instructions are issued. The critical correlates of loss of consciousness is an inhibition of thalamic neurons via reticular-thalamic circuits as measured by increased phase coupling between delta (1 -4 Hz) and gamma (> 20 Hz) frequencies.

Figure 15 shows the changes in cross-frequency synchrony between delta and gamma frequencies during proprofenol induction (loss of consciousness) and recovery (return of consciousness).

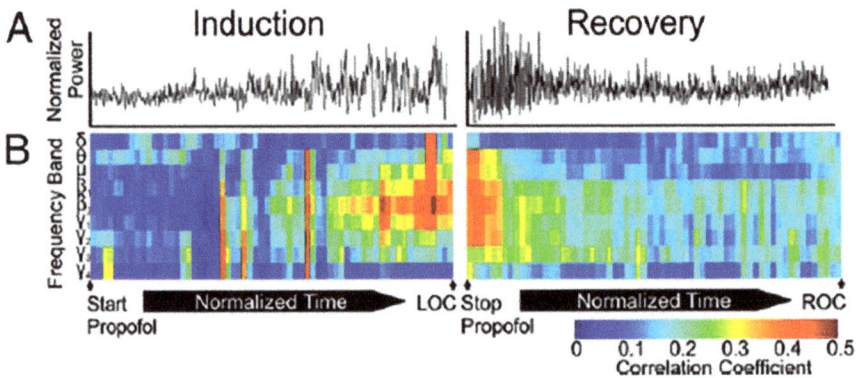

Fig. 15 Trends in μ, β, and γ1 power variability and covariance between distant cortical sites. (A) Average normalized μ, β, and γ1 power from an exemplar

electrode demonstrating increased variance during induction that decreases on recovery. (B) The mean intraband power correlation between electrodes. Power in the μ, β1–2, and γ1 bands shows increasing correlation during induction and a decrease during recovery. δ (1–2 Hz), θ (3–8 Hz), μ (9–11 Hz), β1 (13–21 Hz), β2 (23–35 Hz), γ1 (37–45 Hz), γ2 (75–105 Hz), γ3 (135–165 Hz), and γ4 (195–205 Hz). From Breshearsa et al, 2010.

Importantly, the conscious state is characterized by a frequency shift and a broadening of the width of the spectral distribution that time scale. A remarkable finding was that there was an invariant time scaling between the unconscious state and the conscious state by the ratio = 1.62 or the Golden Proportion. For example, frequency increased by a factor equal to 1.62 ± 0.09 and the change in the width of the spectrum between the U and C states varied by the same ratio (1.61 ± 0.09). The authors accelerated the EEG traces during the unconscious state by an approximate factor of 1.62 and demonstrated that the EEG traces now match the conscious state. There are two stable states operating during consciousness, one is the conscious state and the other the unconscious state and the transition to unconsciousness involves a slowing of the frequency and narrowing the width of the spectrum by the Golden Proportion or 1.62.

Figure 16 shows that during unconsciousness there is a resonant frequency of about 8.2 hz and about 18.2 hz in the conscious state. The conscious state is characterized by a frequency shift and a broadening of the width of the spectral distribution that time scales. A remarkable finding was that there was an invariant time scaling between the unconscious state and the unconscious state and the ratio of the unconscious state

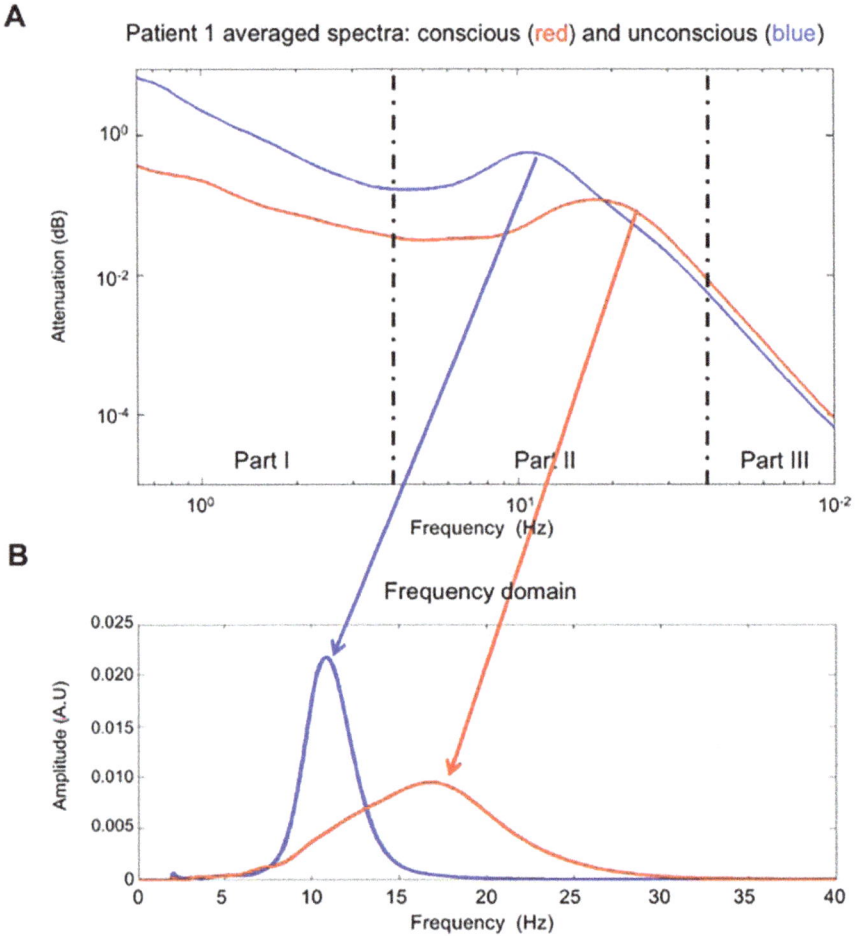

Fig. 16-Typical ECoG frequency and temporal behavior during anesthesia and after recovery. (A) Spectral behavior: log power spectral density during unconsciousness (blue) and after ROC (red). These spectra were obtained from the same EEG channel recorded for 100 sec. in each condition. (B) The spectrum shifts to 18 Hz and broadens when low

frequencies (delta band) decreased power in the 4-40 hz region during unconsciousness (blue) and consciousness. (From Boussen et al, 2018).

and the conscious state is the ratio = 1.62 or the Golden Proportion. For example, frequency increased by a factor equal to 1.62 ± 0.09 and the change in the width of the spectrum between the Unconscious (U) and Conscious (C) states varied by the same ratio (1.61 ± 0.09). The authors shifted the EEG traces during the unconscious state by an approximate factor of 1.62 and demonstrated that the EEG traces then matched the conscious state. This study also showed that there are two stable states operating during consciousness. One is the conscious state, and the other the unconscious state, and the transition to unconsciousness involves a slowing of the frequency and narrowing the width of the spectrum by the Golden Proportion or 1.62.

3
CHAPTER

THE AESTHETIC COMPASS

Aesthetic feeling starts with the subconscious perception of a minimal energy form.

Simplicity and least effort are fundamental to authenticity and genuineness, hence, the aesthetic compass points the pathway in the subconscious mind.

The aesthetic compass is a directional aesthetic feeling. It involves a bottom-up and top down dynamic of human drives and emotions. Action with base emotions such as the urge to move, fear, anger and sexual drives are at the bottom-up dynamic.

The reflective mental appreciation of moments of time and such things as music, art, science and mathematics involve the top-down dynamic. It will be hypothesized that these two compartments of the human psyche, i.e., the "base" and the "aesthetic", represent two dynamically coordinated networks that are phylogenetically different components of the human brain and human behavior. The base emotions are phylogenetically older in their neuroanatomical origins and are largely instinctual and phylogentically invariant from lizards to humans. In contrast, the aesthetic emotions of music, art, science and mathematics involve the neo-cortex and the pre-frontal cortex which is abundant in the human species.

The aesthetic feeling occurs in periods of security and peace when the baser and stronger negative emotions such as fear and anger are absent. Anger and fear turn the aesthetic compass off. Therefore a more stable state is important for aesthetic expression to occur. In mentally healthy individuals aesthetic feeling is a higher emotional feeling, unique to man, when creative and positive visions of the future unfold and the depth of all emotion can be securely and socially expressed. This freedom for creation is a positive force from childhood to old age, for people to create music, rhythm, writings, new thoughts, etc. Music, mathematics, art, athletics, etc. are spatial-temporal forms of action expressed for the pleasure of one self and others. The drive to belong to a group is fundamental to healthy humans and decisions made toward a "more perfect" society and a vibrant economy also include minimal energy judgments. The simpler and more effective a solution, then the greater the reduction in stress and uncertainty, as minimal energy is required for the aesthetic choice.

The positive or negative first meeting with a person is based on trust, which is also influenced by aesthetics. Mentally healthy humans are socially perceptive and the more genuine and authentic a person then the greater is trust toward that person. Simplicity and least effort are fundamental to authenticity and genuineness, hence, the aesthetic compass points the path in the subconscious mind (Jacobsen et al, 2006; Munar et al, 2012

The aesthetic compass of our lives is a subtle and subconscious feeling, referred to as the "aesthetic feeling" (Avram, 2015) that comes immediately and is directionally oriented, which comes effortlessly upon the perception of something beautiful, such as a flower, a sunset, a star, a wave, a sound produced by nature, etc. Each of these elicit or evoke the common feeling of "beauty", which comes to mind immediately and effortlessly without conscious reflection or conscious investigation. The sub-conscious involves bottom-up brain regions that give rise to a resonance with top-down neocortical networks in iterative loops in the brain. The primary aesthetic feeling, however brief in the stream of consciousness is a transformational feeling that has profound and lasting affects on our future actions and future perceptions.

It is in this sense that I align the aesthetic feeling with the concept of an "aesthetic compass". I use the word "compass" as pointing toward the ideal mathematical form of the Golden Proportion sought by our primary sensory systems that approximate a match to the form of certain external forms. These are the forms that give rise to an aesthetic feeling, by virtue of the match itself, resulting in stress relief or reduced psychological tension. This primary aesthetic feeling is a least effort "reward" of emotions such as "joy", "pleasure", "delight", "respect", "awe", "reverence", etc. which approximate the least effort principle in physics that is mathematically expressed as the Golden Proportion because of its pure mathematical minimal energy form.

The Golden Proportion is the only number that is a section of a line such that the smaller part squared equals the larger part which is 1.618033.... or about 2/3 of a line segment.

Simply using this ratio is a huge economy of form! That is, simply duplicate the smaller part to create the whole. There is no other number in the infinite universe of numbers that does this.

The mathematical concept of minimal energy started with the idea of minimal time. The desire of inquisitive people to test these concepts eventually lead to the modern day mathematical physics of Euler and Lagrange and Hamiltonian equations that are direct and fundamental expressions of the concept of minimal energy forms. Plato's Pure "Universe of Mathematical Forms" (UMF), which is the concept of minimal energy forms, is fundamental to the minimal energy property of mathematics. In the case of the mathematics of physics, minimal energy mathematical forms have an important practical and applied value, in contrast to the total Universe of mathematical forms of pure abstract mathematics.

Applied mathematics is a small subset of the Universe of Pure Mathematics, but it shares the same Platonic nature of absolute logical truth and minimal energy forms. By the principals of both classical and operant conditioning each human learns and adjusts behavior and decisions depending on aesthetic feelings of immediate perceptions that approximate minimal energy forms. This is a special attribute of the human primate. The Golden Proportion that gives rise to what I call the "Aesthetic Compass", transcends to also include the other 95% of the Universe called "dark energy" (expanding Universe) and "dark matter" (Black holes). The 5% of the Universe understood by modern physics uses all forces in the Universe that are constrained by the mathematics of thermodynamics, special and general relativity, and quantum mechanics. All of this mathematical physics depends on the concept of minimal energy forms.

3.1 EEG AND AESTHETICS

"The classical frequency bands of the EEG can be described as a geometric series with a ratio (between neighbouring frequencies) of 1.618, which is the golden mean. Here we show that a synchronization of the excitatory phases of two oscillations with frequencies f1 and f2 is impossible (in a mathematical sense) when their ratio equals the golden mean, because their excitatory phases never meet. Thus, in a mathematical sense, the golden mean provides a totally uncoupled ('desynchronized') processing state which most likely reflects a 'resting' brain, which is not involved in selective information processing. However, excitatory phases of the f1-and f2-oscillations occasionally come close enough to coincide in a physiological sense. These coincidences are more frequent, the higher the frequencies f1 and f2. We demonstrate that the pattern of excitatory phase meetings provided by the golden mean as the 'most irrational' number is least frequent andmost irregular. Thus, in a physiological sense, the golden mean provides (i) the highest physiologically possible desynchronized state in the resting brain, (ii) the possibility for spontaneous and most irregular (!) coupling and uncoupling between rhythms and (iii) the opportunity for a transition from resting state to activity." From Abstract Pletzer et al, 2010.

As mentioned in chapter one, while a faculty member at NYU school of medicine I had the good fortune of meeting Eric Swartz and George Chaiken who were trained in mathematics and physics. We spent time examining sensory maps that remarkably obeyed the laws of fluid flow and topology such as conformal maps. As mentioned, the mapping of the retina to the visual cortex is a logrithmic spiral conformal map, the cochlea is a logrithmic spiral, and the mapping of the skin surface to the somatosensory cortex is a logrithmic spiral. Thus, there is a common format across sensory-cortical maps that, because of the mathematical form of the logarithm, and the mathematics of the "Golden Proportion" give rise to computational neuroanatomy by virtue of the mapping itself. The log mapping of the retina to the cortex, for example, provides size and rotational invariance that are intrinsic to conformal mapping. The shape of a computer is irrelevant but the shape of fiber connections and mappings and re-mappings in the brain is a fundamental aspect of brain function. The finding of conformal maps that involve the Golden Proportion suggested a linkage between aesthetic feeling and the form of the sensory-cortical maps themselves. This is significant in brain science because aesthetics is an important part of the human experience and represents a feeling of appreciation of beauty that comes effortlessly and immediately upon contact with certain objects, forms and sounds. The emphasis here is on the concepts of "effortless"and "immediate".

It appears evident that the immediate perception of an object of beauty involves a certain degree of matching between the form of the object (i.e. its proportional properties) and the form of the sensory organization of the brain. Limbic and reticular interactions contribute to the feeling of beauty, the aesthetic components are cortically analyzed, and a "figure of merit" is emerges from the limbic system (e.g., nu. accumbens, amygdala, nu. basalis, etc.). The Golden Proportion, an irrational number 1.618... is in the shape of a snail shell or a sunflower that elicit a limbic-cortical aesthetic figure of merit.

Golden Proportions constitute that class of objects with the least effort transduction by the primary sensory system itself. The ratio of frequencies of the human Electroencephalogram (EEG) also exhibit a Golden Proportion, and in fact this ratio appears to be critical in phase reset dynamics. As seen in figure 17, Pletzer (2010) argued that the Golden Section is the most irrational of all irrational numbers. For example, if one does a Fourier transform of the digits to the left of an irrational

Table 1 – Typical EEG frequency bands and subbands and corresponding periods.					
Frequency band		Frequency subband		Peak	Period
Name	[Hz]	name	[Hz]	[Hz]	[ms]
delta[2]	1.5–4	delta1[3]	1–2	1.5	667
		delta2[3]	2–3	2.5	400
theta[2]	4–10	theta1[*]	3–5	4	250
		theta2[3]	5–8	6.5	154
alpha[1]	8–12	alpha[3]	8–12	10	100
beta[2]	10–30	beta1[3]	12–20	16	62.5
		beta2[3]	20–30	25	40
gamma[2]	30–80	gamma1[3]	30–50	40	25
		gamma2[3]	50–80	65	15
fast[2]	80–200	ripples1[*]	80–120	100	10
ripples[4]		ripples2[*]	120–200	160	6.25

Fig. 17- 3.5–4 Hz) at the lower limit of the broad theta band would complete the geometric series with ratios theta1/delta2 = 4/2.5=1.6 and theta2/theta1=1.625. This would result in two stable subbands of theta, as has been demonstrated for the beta and gamma band in humans and for the delta band in vitro (see Roopun et al., 2008a,b). Note: Typical harmonic relationships for slow frequencies: Delta/Theta/Upper Alpha=3:6:12,

Typical harmonic relationships for high frequencies: Upper Alpha/Beta/Gamma=12:24:36:48. 1 First EEG rhythm described in humans (Berger, 1929). 2 Rhythms reported by Buzsaki and Draguhn (2004) from multiple recordings in mice, rats and humans. 3 Stable rhythms generated in vitro by Roopun et al. (2008a,b). 4 High frequency oscillations as described in human epileptogenesis (see, e.g., Zelman et al., 2009). From Pletzer et al, 2010

number (like the square root of 2 or square root of 7) then one finds repeating segments or sequences of numbers giving rise to peaks in the spectrogram.

In contrast the Fourier transform of the digits to the right of the decimal point of the Golden Proportion irrational number is nearly flat and without peaks. Experiments by Peltzer et al (2010) demonstrated that phase shifts dominate the Golden Ratio of EEG frequencies and the probability of phase lock is at a minimum. Thus, complexity and effortless dynamics are intrinsic to the cross-frequency coupling of EEG rhythms.

The relationship between aesthetic feelings and "complexity" is often defined in terms of "perceptual effort". This linkage between the complexity of an object and perceptual effort is paralleled by a similar linkage between mathematics and the simplicity of the physical laws of the universe. For example, the complex logarithm function, which is characteristic of the global and local structure of the sensory mappings of the brain, may also be used to describe the pattern of electric or magnetic fields, the velocity flow of a fluid, or the distribution of a diffusing chemical reactant. The basic developmental reason for this commonality is that structures of these sorts require minimal encoding. That is, they represent the most parsimonious and economical methods of controlling dynamic flow, and in the case of living matter, the growth of form. The unifying and simple nature of these observations indicates that their commonality is ultimately an expression of the "Variational Principle" or the principle of "Least Effort" in physics. This principle, as reflected in the calculus of variations, is a description of the processes by which nature finds the path of least resistance, or the solution of elegance, simplicity and parsimony in the resolution of conflicting forces in nature (a sunset, a flower or a tornado and hurricane). The mechanisms of aesthetics and perhaps more generally, perception, may directly involve a similar linkage. In this case the least effort expression for the growth of living forms (the Golden Proportion) is matched by the least effort expression for the evolution of the physical universe. In other words, a fundamental aspect of the human aesthetic feeling involves a match between the organizational laws of the atoms of the brain and the organizational laws of the atoms of the environment or space external to the brain. The mathematical variational principle of Euler-Lagrange and Hamilton is a universal expression that applies to living matter and human consciousness. In the case of human consciousness, the negative entropy of the synchronous discharge of millions of neurons is time linked to the present. This is matched and mismatched to memory and expectations of the future in 80 to 300 millisecond intervals of time, and are sequential minimal energy states. The fundamental neural process of effortless phase resetting and phase realignment of billions of neurons occurs with no net expenditure of energy and thus the high speed recruitment of large numbers of neurons is itself a minimal energy form (Thatcher, 2016).

3.2 GOLDEN SECTION AND THE VARIATIONAL PRINCIPLE

The preceding discussion provides a basis for a discussion of the role of the Golden Proportion in what may be called the postulates of an "economy of complexity". We must first, however, define "economy". The Webster dictionary defines it as "The careful or thrifty management of resources". Oddly enough, a very similar definition applies to physics and the laws of nature. For example, in physics, economy is defined by the "variational principle" or the "principle of least effort". This is a principle of nature by which the laws of physics follow the path of least energy or least effort. Pythagoras (c. 530 B.C.) introduced this concept when he revealed fundamental laws of acoustics. Aristotle (384 -322 B.C.) advanced this by introducing the concepts of simplicity, economy and least effort into natural science through the application of a "minimum hypothesis" requirement for establishing scientific truth, which also required hypotheses to be "simple" and "primary". Following Ockham (c. 1300 -1347), Kepler (1571 -1630) postulated a "metaphysics" based on the principle of economy, simplicity and least effort, in which the universe was "a mathematico-aesthetic numerical harmony and exhibiting a surpassing simplicity and unity -"natura simplicitatem amat". The mathematical application of these concepts of "simplicity", "economy", and "least effort" to physics was later formalized by Leibniz (1646 -1716) when the concept of "economy" became a postulate which acts as a foundation for the development of "minimum" and "maximum" concepts in calculus. Leibniz conceptualized this on a personal level when he stated: "

"The perfectly acting being . . . can be compared to a clever engineer" who obtains his effect in the simplest manner one can choose".

There are many examples of a process of "economy of effort" found in nature. This fact, in which nature appears to do things with the greatest economy of effort, led some scientists and philosophers to attach religious significance to this principle and to extend their investigations into the realm of mathematics. In this vein Leonard Euler, an eighteenth century mathematician (1707 -1783), developed some fundamental concepts of a branch of mathematics known as the "calculus of variations". In this discipline, limits are placed on a function so as to provide an extreme value for an integral involving the function itself. With this method the maximum or minimum area or volume of a function (actually the sum of differences) can be evaluated over intervals of time that later turned out to have great importance in thermodynamics and relativity theory. In 1788 Lagrange adapted the calculus of variations to problems in the dynamics of bodies in motion and produced some fundamental results. A climax to this work was reached in 1834 when Hamilton announced the principle which now bears his name (i.e., "The Hamiltonian") and which occupies a unique place in all of science for its elegance and simplicity. The Euler-Lagrange equation and the Hamiltonian equation are eloquent and succinct descriptions of the operation of "economy of effort" in nature and form a fundamental part of modern physics including Relativity Theory, without these, man would not have reached the moon or developed microwave ovens or computers. In the case of aesthetics, iterative loops in the brain structured by the log spiral and Golden Proportion minimize the distance from the ideal and involve regression toward the most efficient outcome at each moment of time.

3.3 MATCH-MISMATCH OF "MINIMAL ENERGY FORMS" IS AN AESTHETIC COMPASS

As mentioned previously, the concept of an aesthetic compass suggests a universal "guide" or "direction" to the emotion of "aesthetic feeling". The initial component of aesthetic feeling arises from a match between the physical structure of the brain to the "ideal" mathematical structure of the universe. It seems astounding to suggest that any such unifying or simple "guide" may exist within the mathematical rules of the Universe which simultaneously governs the physical laws of the brain, yet this logically follows. It will be postulated in the paragraphs to follow that: Immediate reward and pursuit of the aesthetic feeling arises from a least effort minimal energy match between the form of an external stimulus and the form of the mapping of that external stimulus into the brain. This is another example of mathematical physics using the principal of least action or least effort and minimal energy.

In order to explore this possibility, one fundamental questions must first be answered: Why is there a Golden Proportion minimal energy mapping of the sensory receptors to the neocortex of the human brain? A second question, based on the validity of the first question: Is the reward and pursuit of the aesthetic feeling based upon a minimal energy matching due to this minimal energy mapping of sensory receptors to the neocortex of the brain? For the moment, let us accept the science reviewed previously in which the neurophysiological mapping of the retina to the cortex is shown to be a logarithmic spiral map like a snail shell; and similarly, that the somatosensory and auditory systems involve the same logarithmic spiral map. Further, accept that physics tells us that such mappings are minimal energy forms and this is in fact a common format for major senses. Why is this true? What is the survival value of a Golden Proportion mapping from the lowest level of sensory receptors to the highest level of the neocortex?

I would like to take us back to the simpler times of the ancient Egyptians and Greeks and once again consider "rhythm" in the context of "proportion" and "number". For example, the notions of periodicity and proportion, and their interactions, can be used to describe succession in time as well as organizations in space. As was defined by the ancient Egyptians, "rhythm is in time what symmetry is in space". That is, if periodicity is the characteristic of rhythm in time, then proportion is the characteristic of rhythm in space. Combinations of proportions can bring periodic reappearances of proportions and shapes, just as in a musical chord or the successive notes or chords of a melody we may perceive an interplay of proportions. It was this unification of the concepts of proportion, rhythm and symmetry that led Pythagoras to express consonant musical intervals by simple ratios, i.e. by fractions whose numerator and denominator are members of the "*tetractys*", series or 1,2,3,4. Pythagoreans discovered that one could produce "pleasing" musical intervals by dividing a vibrating string in ratios such that the ratio 1:2 yields an octave, 3:2 yields the fifth, and 4:3 yields the fourth. Subsequent work in the aesthetics of music has revealed other aesthetically pleasing ratios such as the major sixth (i.e. 8/5). In summary, we can equate through mathematics, proportions in time, and rhythm in space, to proportions in space and rhythm in time.

3.4 THE EVOLUTIONARY VALUE OF THE GOLDEN PROPORTION

Speculation about the brain mechanisms of aesthetics involves traveling in rechartered waters. We might ask "Why do humans experience feelings of aesthetic appreciation"? In a more biological vein "What is the survival value of the feeling of aesthetics"? A partial answer can be found in the notion that homo sapiens are distinguished from other animals by their relative independence from environmental niches and by their creativity. For example, the desert fox is intimately subservient to its own niche with only limited abilities to alter the environment. In contrast, homo sapiens have developed an expanded central nervous system which provides a type of general purpose computer by which humans have an enormous (but not unlimited) ability to modify and manipulate the environment to suit their needs. However, humans, with their enhanced brain/body ratio and dexterous hand and thumb came into this world without explicit instructions or an "operating manual" that tells an individual how best to behave in a given situation. Therefore, we might ask: How do individuals make the correct decisions needed to constructively alter their environment? How does man know that the elements at his disposal, when placed in the right ratio and proportion, will yield a correct solution to his problems? The answer may be that when there is an approximate match between the "least effort" physical or mathematical laws governing growth and development in the universe and the laws governing the growth and development of the brain, then an aesthetic feeling is made possible. The creative, idealistic and most important, the "corrective" processes by which man builds and develops his life and environment are often guided by aesthetic feelings.

This should not be taken to mean that only the forms of the golden proportion elicit aesthetic feelings. But, the approximation of this form often may be part of the process involved in the elicitation of an aesthetic experience.

4

CHAPTER

MODELS OF AESTHETICS

Philosophers, theologians, and writers on aesthetics have been attempting to formulate the bases for aesthetics for more than 2,000 years. A contemporary mathematical model of aesthetics was formulated by George Birkhoff in 1933. Birkhoff, in the same vein as Pythagoras and other Greek philosophers, contends that the attributes upon which aesthetic value depends are accessible to measure. For example, Birkhoff argued that the aesthetic experience is made up of three successive phases:

> "(1) a preliminary effort of attention, which is necessary for the act of perception, and which increases in proportion to what we shall call the complexity (C) of the object; (2) the feeling of value or aesthetic measure (M) which rewards this effort; and finally (3) a realization that the object is characterized by a certain harmony, symmetry,

or order (O), more or less concealed, which seems
necessary to the aesthetic effect."

These are combined in the basic formula:
M = O/C

which expresses the hypothesis that the aesthetic measure is determined "by the density of order relations in the aesthetic object." Birkhoff proceeded from this basic definition to a consideration of what appeals to us in polygonal forms, ornaments, vases, diatonic chords and harmony, melody, and the musical quality in poetry, emphasizing that they are inter-related. Birkhoff supplies formulas enabling values of C and O to be computed for polygons, vase outlines, melodies, and lines of poetry. Unfortunately, the governing principles are peculiar to each class of material and no general rule or principle is developed. In polygons, C is the number of indefinitely extended straight lines that contain all the sides, and O increases with the properties of symmetry and horizontal-vertical orientation. In vases, C is identified with a number of "characteristic points" on which "the eyes can rest". In music, C is a number of notes in a melody, and O depends on melodic harmonic sequences. In poetry, C is the number of elementary sounds plus the number of word junctures that "do not admit of liaison," and O is derived from the presence of rhyme, alliteration, and assonance. Birkhoff did not develop a general principle for the concepts of harmony, symmetry or order, nor did he enter some of the classical concepts such as "analogia" into the equation. Further, this model indicates that the simplest and most regular patterns will be highest in aesthetic value. This is not, however, consistent with the asymmetry of the golden proportion or with the results of experiments in which subjects have been required to judge polygons selected from those illustrated in Birkhoff's book (Davis, 1936; Eysenc, 1941).

Berlyne (1971) presented an interesting information-theoretic modification of the Birkhoff equation. According to this modification C is identified with uncertainty (H) and O with redundancy (R = (Hmax -H)/Hmax). By substituting these values in the formula M = O/C, it is inferred that M = 1/H -1/Hmax. Unfortunately, this formulation does nothing to remove the earlier criticisms against the Birkhoff model.

A neurophysiologically based mathematical model of aesthetics was proposed by Rashevsky in (1938). The neurophysiological principles were speculative but emphasized the role of reciprocal excitation and inhibition within populations of neurons. Aesthetic value depended on the "total net excitation" that is transmitted to a "pleasure center" (location unspecified). Rashevsky presented the interesting argument that symmetry reduces "effective complexity" and thus it reduces excitation " by subtraction rather than division, i.e., by adding inhibition". In other words, it is possible for a slightly asymmetrical form to have greater aesthetic value than a purely symmetrical form like a square. This relationship suggested to Eysenc (1942) that the equation would be most in keeping with experimental data presented by Davis (1936) and best represents the relationship

$$M = O\ C$$

between order (e.g. repetition, sequence, symmetry) and complexity (effort of attention). According to this equation, aesthetic value can be increased by increases in either complexity or order. However, these two measures are to some extent inversely related and depend on the precise scales used. A serious problem with a multiplicative formulation is that both order and complexity are directly proportional to aesthetic measure. Thus, as complexity decreases, aesthetic measure decreases. Such a relationship is not consistent with the form of the golden proportion or with modern experiments that test aesthetic measure (Svensson, 1977; Benjafield, 1976; Benjafield and Green,1976 and Benjafield et al, 1980).

4.1 COMPLEXITY AND INFORMATION THEORY

Information theory grew out of the work of Wiener (1948) on 'cybernetics' (defined as the science of control and communication) and of Shannon (Shannon and Weaver, 1949) on the 'mathematical theory of communication'. It has subsequently been used for the solution of a wide range of problems in the area of communication and concepts of order.

Information theory begins at the level of "difference" by treating all patterns or events in space or time as 'bits' of information. A bit is a basic unit of $P = -\sum = p_i \log_2 p_i$ "difference" and information theory represents one method by which orders of bits can be distinguished and quantified. This formalism was accomplished by first introducing the concept of "uncertainty" which can be quantified whenever we have a "sample space" of items (signals embedded in noise). Uncertainty arises in situations in which some event is to be selected from a set of alternative classes of events. We may not know which event will occur, but we can enumerate the alternative classes and assign a probability to each of them. The value of uncertainty, in bits, is provided by Shannon's famous formula, , in which pi is the probability that the event will belong to class i. Defined in this way, uncertainty has two important properties: it increases with the number of alternative classes of events; and, if the number of alternative classes is held constant, it reaches a maximum when the classes are equally probable.

In terms of aesthetics it can be argued that every element of color or form or shape, etc, is a particular set that is selected from a larger set of alternatives and can be regarded as signals. However, for any particular art form and style, the set from which each element is selected is limited. The alternatives that can occur in a particular location constitute a sample space. Their relative frequencies can be calculated and a probability associated with each of them. Consequently, every location in a work of art, whether spatial or temporal, can be allotted an uncertainty value.

The relationship between uncertainty and information can be understood once the awaited signal has appeared and we know which alternative has been chosen. At this point we can assign an "amount of information". This assignment, which is measured in bits, will be greater, the lower the probability of the class to which the signal belongs, the appropriate formula being: $-\log_2 P_i$.

In accordance with this formula, the amount of information varies between zero and 'infinity' as the choice of the event in question varies between certainty and impossibility. In this regard, uncertainty can be equated with the average or expected amount of information, which can be calculated before the choice is revealed, whereas the actual amount of information cannot be specified until it is clear which choice was made.

The concept of complexity pertains to art, music and mathematics when it is defined by the "limited channel capacity" of the human sensory systems. As is well documented in the cognitive psychology literature, human information processing has very clear and definable limits, with the channel capacity of approximately 7± 2 (Miller, 1956) items per unit presentation. Thus, both simultaneous and successive aspects of information must be considered in understanding the perceptual capacity of the human nervous system. One way to accomplish this is to use the equation

$$K \log_2 x$$

to represent information conveyed by a work of art in the form of ordered temporal and spatial 'bits'. According to this equation, K log2 x is a metric of complexity, where K is a constant whose value depends on how many successive elements are perceived per unit time, and log2x is the number of independent 0-or-1 choices represented simultaneously, that is, the number of independent features that can be perceived at once. Thus, "complexity" is related to the number of 'bits' transmitted both simultaneously and serially as an individual perceives an object or thought.

4.2 UNCERTANTY REDUCTION AND EFFICIENCY MODEL OF AESTHETICS

Early models of aesthetics did not benefit from modern neurophysiological data showing that the primary sensory surfaces of the body map to the cortex by a logarithmic conformal mapping. Consequentially no emphasis has been placed on the role of sensory-cortical maps in the production of aesthetic feeling. Nor is there an understanding of the active process of perception in which models of reality are continuously being compared (match-mismatch) to sensory input. In an improved formulation "perceptual effort" as used in Birkoff's theory I add the concepts of complexity and information theory and a dynamic oscillation around zero by the human brain. Birkhoff defined complexity as "a preliminary effort of attention, which is necessary for the act of perception and which increases in proportion to what we shall call "complexity". A modification is to state that perceptual effort does not necessarily increase in direct proportion to the complexity of an object. That is, the perception of an object requires a mapping process whereby the form of the object is transformed through a conformal map, and primary emphasis should be placed on the result of that transform or the "internal representation" of the object, as a match and mismatch of an ideal mathematical form with respect to a separate Universe of ideal forms. The brief time gap between the separate Platonic Universe of Ideal Forms and human perception is "Quantum Mechanics" by the metric of Planc's constant = 10^{-43} seconds.

Forms which approximate the golden proportion and thus the logarithmic spiral mapping of the brain should result in the least amount of perceptual effort by virtue of the "fit" of external to internal structure. One would expect a "resonance" or "match" between the least effort atomic form of the external world with the least effort atomic form of the brain. That is, the anatomical and neurophysiological sensory maps are spatially and temporally organized "atoms" and the organizational form of these atoms is in the form of the mathematically ideal of the "Golden Proportion". The least effort organization of the physics of atoms is described by the Euler-Lagrange equations, and the Hamiltonian equation lead to "concentric circles,", "radial lines" and "logarithmic spirals" in the description of cosmic processes such as "Galaxies", "Black Holes", the weather such as "Tornadoes and Hurricanes", or simple household process such as a "Sink Draining". The Golden Proportion is indeed ubiquitous in nature.

This would explain why a asymmetrical form such as the golden proportion is more aesthetically pleasing than a symmetrical form and why the formula M = O/C is deficient in predicting that the simplest and most regular patterns have the highest aesthetic value. Based on these considerations the following formula is posited:

A = [D]/C

Where A is an "aesthetic measure", C is a measure of complexity and |D| is the absolute distance between a template match of the golden proportion in its ideal form and the form of a given sensory-cortical map (or |X phi -X object| > 0). This relationship can be more precisely described in information theoretical terms:

M = $g(\theta)$/(Klog2x)

Where:

$$g(\varphi) = \int_{-\infty}^{+\infty} f(\varphi) h(\varphi - \delta) d\varphi$$

The numerator of the equation is a cross-correlation function (or convolution) representing the degree of match of an form $f(\theta)$ external to the body with the logarithmic spiral form $h(\theta)$ in the sensory mappings of the brain. The output of the cross-correlation function $g(\theta)$ is in terms of neurophysiological excitation (e.g. increase in neuronal discharge rate) which exhibits a form of a "tuning curve", where the greater the approximation of the external form to the golden proportion then the greater the neurological output. The denominator of the equation (K log2 x) is a metric of complexity (since it is the inverse of complexity it is actually a measure of "simplicity"), where K is a constant whose value depends on how many successive elements are perceived per unit time, and (log2 x) is the number of independent 0-or-1 choices represented simultaneously or the number of independent features that can be perceived at once. The ratio of these two functions (the cross-correlation with the brain divided by the complexity of an object) yields the aesthetic measure of the object. In engineering terms, the log spiral mapping of the body surface (actually the sensory transducers) onto the primary sensory cortex represents a type of a "transfer function" $h(\theta)$, with the input $f(\theta)$ coming from the external world and the output $g(\theta)$ coming from the primary sensory cortex and then projecting to secondary sensory cortex as well as to other regions of the brain.

According to this formula, aesthetic measure is directly proportional to the classical measures of aesthetics; namely: proportion, analogia, order and symmetry while being inversely proportional to complexity. The denominator of the equation (K log2 x) represents the magnitude of both order (K) and complexity (log2 x). The central concept of "economy of complexity" is represented through the ratio of a metric of order and complexity of an object, or form in the external world, to the degree of match that this form has with the golden proportion in the central nervous system.

Forms which are maximally simple and maximally matching (with the Golden Proportion form of the brain) will yield a maximum aesthetic value. These concepts are summarized below.

MATHEMATICAL PRINCIPLES:

1. Fibonacci Series: An economy of predicting the future based on the past.
2. Golden Section: An economy of geometric asymmetry.
3. Golden Proportion: An economy of form.
4. Logarithmic Spiral: An economy of complex growth and form.

NEURO-SENSATION PRINCIPLES:

1. Mapping of retina to the neocortex follows a logarithmic spiral form.
2. Mapping of skin surface to the neocortex follows a logarithmic spiral form.
3. Mapping of sound frequency to the cortex follows a logarithmic form.

AESTHETIC PRINCIPLES:

1. An aesthetic feeling involves an immediate and effortless recognition of beauty.
2. Aesthetic feeling is a continuum reaching its maximum when the degree of matching of the external form with the golden logarithmic mapping reaches its maximum.
3. Aesthetic feeling is the product of the complexity of an object and the degree of match of that object with

the Platonic mathematical Universe of ideal form in the brain.

4. The mathematical formula where M is the aesthstric measure is: $M = g(\theta)/(K \log 2x)$.

where the numerator of the equation is a cross-correlation function (or convolution)representing the degree of match of an external form f(j) with the logarithmic spiral form h (j) in the brain. The output of the cross-correlation function g(j) is in terms of neurophysiological excitation (e.g., increase in neuronal discharge rate) which exhibits a form of a "tuning curve", where the greater the approximation of the external form to the golden proportion then the greater the neurological output. The denominator of the equation (K log2x) is a metric of complexity (since it is the inverse of complexity it is actually a measure of "simplicity"), where K is a constant whose value depends on how many successive elements are perceived per unit time, and (log2x) is the number of independent 0-or-1 choices represented or perceived simultaneously. The ratio of these two functions (the cross-correlation with the brain divided by the complexity of an object) yields the aesthetic measure M of the object. In engineering terms, the log spiral mapping of the body surface onto the primary sensory cortex represents a type of a "transfer function" h(j) with the input f(j) from the external world and the output g(j) coming from the primary sensory cortex and projecting to secondary sensory cortex as well as to other regions of the brain (e.g., limbic and thalamus). According to this formula: *"forms which are maximally simple and maximally matching are maximally aesthetic".*

4.3 AESTHETICS OF ABSTRACTIONS AND VARIATION

As defined in the introduction, the measure of aesthetics that we are concerned with represents a feeling of appreciation of beauty which comes effortlessly and immediately upon perception of certain objects and sounds. The emphasis here is on the concepts of "effortless"and "immediate". It appears evident that the immediate perception of an object of beauty involves a certain degree of matching between the form of the object (i.e. its proportional properties) and the form of the sensory organization of the brain. Although it is beyond the scope of the present paper to explore detailed limbic and reticular contributions to the awareness and feelings of beauty, it seems that the aesthetic components are cortically analyzed and a "figure of merit" is assigned by the limbic system (e.g., nu. accumbens, amygdala, nu. basalis, etc.). This limbic-cortical aesthetic figure of merit is chemically rewarded by the perceptions of the golden proportion because golden proportions constitute that class of objects with the least effort transduction by the primary sensory system itself. An extension or elaboration of of these concepts can be achieved to account for higher and abstract levels of aesthetic feeling. For example, if one assumes that matching to memory and logical operations also involve brain anatomical mappings that follow PHI, then a general integral can be written

$$M_f = \int_0^{\tau} a_1 M_1 + a_2 M_2 + a_3 M_3 + \iota\ a_n M_n$$

in which Mf is the final aesthetic measure, which is the sum of serial stages of cognitive processing, each of which involves a mapping process that contains the golden proportion and a specific weighting *an* . This linear model can be modified to include nonlinear feedback between early stage and later stage outputs, thus resulting in a dynamic aesthetic filter process in which past experiences and current passions weight the aesthetic value of an experience in a given moment of time. In this way, a general model can be developed to explain secondary aesthetic feelings which require learning and memory processes.

5

CHAPTER

REFERENCES

Agnati LF, Agnati A, Mora F, Fuxe K (2007) Does the human brain have unique genetically determined networks coding logical and ethical principles and aesthetics? From Plato to novel mirror networks. Brain Res Brain Res Rev 55(1):68–77.

Ahlfors, L. (1966). "Complex Analysis", New York: McGraw Hill.

Allan, L.G. (1978). Comments on current ratio-setting models for time perception. "Perception and Psychophysics", 24,444 -450.

Allman, J.M. and Kaas, J.H. (1974). The organization of the second visual area (V-II) in the owl monkey: A second order transformation of the visual hemifield. Brain Research; 76, 247 -265.

Avram, M., Gutyrchik, E., Bao, Y., Pöppel, E., Reiser, M., & Blautzik, J. (2013). Neurofunctional correlates of esthetic and moral judgments. Neuroscience Letters, 534, 128–132. doi:10.1016/j.neulet.2012.11.053

Benjafield, J. (1976). The 'golden rectangle': Some new data. "Amer. J. Psychol." 89, 737 -743.

Benjafield, J. and Adams-Webber, J. (1976). The golden section hypothesis. Br. J.Psychol., 67, 11 -15.

Benjafield, J. and Green, T.R.G. (1978). Golden section relations in interpersonal judgement. Br. J. Psychol., 69, 25 -35.

Benjafield, J., Pomeroy, E. and Saunders, M. (1980). The golden section and the accuracy with which proportions are drawn. Canad. J. Psychol. Rev. Canad. Psychol., 34, 253 -256.

Berlyne, D.E. (1971). Aesthetics and Psychobiology, Appleton-Century-Crofts: New York.

Berridge, K.C. and Kringelbach, M.L. (2015). Pleasure systems in the brain. Neuron. 86(3): 646–664. doi:10.1016/j. neuron.2015.02.018.

Birkhoff, G.D. (1933). "Aesthetic Measure", Cambridge, Mass: Harvard Univ. Press. Breshearsa, J.D., Roland, J.L., Sharma, M., Gaona, C.M, Freudenburg, Z.V., Templehoff, R.,

Avidane, M.S. and Leuthardt, E.C. (2010). Stable and dynamic cortical electrophysiology of induction and emergence with propofol anestheia. PNAS, 107(49): 21170-21175.

Brown S, Gao X, Tisdelle L, Eickhoff SB, Liotti M (2011) Naturalizing aesthetics: Brain areas for aesthetic appraisal across sensory modalities. Neuroimage 58(1):250–258.

Bohm, D. (1969). Some remarks on the notion of order. In: C.H. Waddington (Ed.), "Towards a Theoretical Biology. II", Aldine Pub. Co, Chicago, pp. 18-58.

Castro D.C. and Berridge K.C. (2014). Opioid hedonic hotspot in nucleus accumbens shell: Mu, delta, and dappa maps for enhancement of sweetness "liking" and "wanting". J Neurosci.; 34:4239– 4250. [PubMed: 24647944]

Cavanagh, P. (1978). Size and position invariance in the visual system. "Perception", 7, 167 -177.

Cela-Conde C.J., et al. (2018) Activation of the prefrontal cortex in the human visual aesthetic perception. Proc NatCela-Conde l Acad Sci USA 101(16):6321–6325.

Chatterjee A., Vartanian, O. Ann (2016). Neuroscience of aesthetics. N Y Acad Sci.,1369(1):172-94. doi: 10.1111/nyas.13035. Epub 2016 Apr 1.PMID: 27037898

Cooper, J. M. and Hutchinson, D.S., eds. (1997). *Plato: Complete Works*. Hackett Publishing.

Daniel, P.M., and Whitteridge, D. (1961). The representation of the visual field on the cerebral cortex in monkeys. J. Physiology, 159, 203 -221.

Davis, R.C. (1936). An evaluation and test of Birkhoff's aesthetic measure and formula. J. Gen. Psychol., 15: 231-240.

Efron, E. (1967). The duration of the present. Annals of the New York Academy of Sciences, 138,713-729.

Efron, R. (1970a). The relationship between the duration of a stimulus and the duration of a perception. Neuropsychologia, 8, 37-55. (a)

Efron, R. (1970b). The minimum duration of a perception. Neuropsychologia, 1970,8, 57-63. (b) Eysenc, H.J. (1941). The empirical determination of an aesthetic formula." Psycho. Rev; 31, 83 -92.

Eysenc, H.J. (1942). The experimental study of the "good Gestalt" -A new approach. Psych. Review, 49, 344 -364.

Fox, P.T. (2005). The human brain is intrinsically organized into dynamic, anticorrelated functional networks. Proc. Natl Acad. Sci. USA 102, 9673–9678.

Ghyka, M. (1977). "The Geometry of Art and Life", New York: Dover Publications.

Hebb, D. O. (1940). "Human Behavior After Extensive Bilateral Removal from the Frontal Lobes". Archives of Neurology and Psychiatry 44 (2): 421–438. doi:10.1001/archneurpsyc.1940.02280080181011

Honrubia, V. and Ward, P.H. (1968). Longitudinal distribution of the cochlear microphonics inside the Cochlear Duct (Guinea Pig). J. Acoust. Soc. Am., 44, 951 -958.

Hofstader, D.R. (1980). "Godel, Escher, Bach: An Eternal Golden Braid", Basic Books, New York.

Hubel, D.H., and Wiesel, T.N. (1962). Receptive fields, binocular interaction and functional architecture in the cats visual cortex. J. Physiology, 160, 106 -154.

Hubel, D.H.,and Wiesel, T.N. (1974). Sequence regularity and geometry of orientation columns in the monkey striate cortex. J. Comparative Neurology, 158, 267 -293.

Hubel, D.H. and Livingstone, M. (1981). Regions of poor orientation tuning coincide with patches of cytochrome oxidase staining in monkey striate cortex. Neuroscience Abstracts, 7, 357.

Huntley, H.E. (1970). "The Divine Proportion", New York: Dover Publications.

Jacobsen T., Schubotz RI, Höfel L, and Cramon, D.Y. (2006). Brain correlates of aesthetic judgment of beauty. Neuroimage 29(1):276–285.

John, E.R. (2005). From synchronous neural discharges to subjective awareness? Progress in Brain Research, Vol. 150: 143-171.

Kaplan S (1987) Aesthetics, affect, and cognition. Environ Behav 19:3–32.

Kawabata, H., and Zeki, S. (2004). Neural correlates of beauty, Journal of Neurophysiology 9: 1699– 1705.

Lacey S, et al. (2011) Art for reward's sake: Visual art recruits the ventral striatum. Neuroimage 55(1):420–433.

Lee U, Mashour GA, Kim S, Noh GJ, Choi BM. (2009). Propofol induction reduces the capacity for neural information integration: implications for the mechanism of consciousness and general anesthesia. Conscious Cogn. 18(1):56–64. PMID 19054696 doi:10.1016/j.concog.2008.10.005

Mandelbrot, B.B. (1982). The fractal geometry of nature. W.H. Freeman, San Francisco.

Merzenich, M.M., Knight, P.L., and Roth, G.L. (1975). Representation of the cochlea within primary auditory cortex of the cat. J. Neurophysiology, 231 -249.

Merzenich, M.M., Kaas, J.H. and Roth, G.L. (1976). Auditory cortex in the grey squirrel; Tonotopic organization and architectonic fields. J. Comparative Neurology, 166, 387 -402.

Merzenich, M.M. and Brugger, J.F. (1973). Representation of the cochlear partition on the superior temporal plane of the macaque monkey. Brain Research, 50, 275 -296.

Miller, G.A. (1956). The magical number seven, plus or minus two: some limits on our capacity for processing information. Psychol. Rev., 63, 81 -97.

Mountcastle, V. (1957). Modality and topographic properties of single neurons of cats somatic sensory cortex. J. Neurophysiology, 20, 408 -434.

Munar E, et al. (2012) Lateral orbitofrontal cortex involvement in initial negative aesthetic impression formation. PLoS ONE 7(6):e38152.

Nadal, M., Munar E., Capó M.A., Rosselló J., Cela-Conde C.J. (2008) Towards a framework for the study of the neural correlates of aesthetic preference. Spat Vis 21(3–5):379–396.

Nadal M. and Chatterjee A. (2019). Neuroaesthetics and art's diversity and universality. Wiley Interdiscip Rev Cogn Sci.,10(3):e1487. doi: 10.1002/wcs.1487. Epub 2018 Nov 28.PMID: 30485700

Penrose, R. (2005) "The Road to Reality: A Complete Guide to the Laws of the Universe", Oxford Univ. press.

Phillips F, Norman JF, Beers AM. Fechner's aesthetics revisited (2010). Seeing Perceiving. 23(3):263-71. doi: 10.1163/187847510X516412.PMID: 20819476

Piaget, J. (1975). Biology and Knowledge. Chicago: University of Chicago Press (2nd Edition).

Pletzer, B., Kerschbaum, H., and Klimesch, W. (2010). When frequencies never synchronize: The golden mean and the resting EEG. Brain Research, 1335: 91-102.

Poggio, G.F. and Fischer, B. (1977). Binocular interaction and depth sensitivity in striate and pre-striate cortex of behaving rhesus monkey. J. of Neurophysiology, 40, 1392 -1405.

Polyak, S. (1941). "The Retina, Chicago": University of Chicago Press.

Rabinovich. M.I., Afraimovich, V.S. Christian, B. and Varona, P. (2012). Information flow dynamics in the brain, Physics of Life Reviews 9: 51–73.

Raichle ME, et al. (2001) A default mode of brain function. Proc Natl Acad Sci USA 98(2):676–682.

Rashevsky, N. (1938). Contribution to the mathematical biophysics of visual perception with special reference to the theory of aesthetic values of geometrical patterns. Psychometrika, 3, 253 -271.

Romani, G.L., Williamson, S.J. and Kaufman, L. (1982). Tonotopic organization of the human auditory cortex. Science, 216, 1339 -1340.

Salimpoor, V.N, van den Bosch,I., Kovacevic,N., McIntosh, A.R., Dagher,A. and Zatorre, R.J. (2013). Interactions Between the Nucleus Accumbens and Auditory Cortices Predict Music Reward Value. Science, 340: 216-219 DOI: 10.1126/ science.1231059 2013

Schwartz, E.L., (1977a). Spatial mapping in primate sensory projection and relevance to perception, Biological Cybernetics, 25, 181 -194.

Schwartz, E.L. (1977b). Afferent geometry in primate visual cortex and the generation of neural trigger features, Biological Cybernetics, 69, 655 -683.

Schwartz, E.L. (1980). Computational anatomy and functional architecture of striate cortex: A spatial mapping approach to perceptual coding, Vision Research, 20, 645 -669.

Schwartz, E.L. (1984). Spatial mapping and spatial vision in primate striate and infero-temporal cortex, In: L. Spillman and B. Wooten (EDs), Sensory Experience, Adaptation, and Perception: A Festchrift for Ivo Kohler, Erlbaum Assoc., Hillsdale, N.J., pp. 73 -104.

Schwartz, E.L. (1985). On the mathematical structure of the visuotopic mapping of Macaque striate cortex. Science, 227, 1065 -1066.

Shannon, C.E. and Weaver, W. (1949). "The Mathematical Theory of Communication", Chicago Univ. Press, Urbana, Ill., 1949.

Somjen, G. (1972). "Sensory Coding in the Mammalian Nervous System", New York: Appleton-Century-Crofts.

Strykker, M., Hubel, D.H. and Wiesel, T.N. (1977). Ann. Meeting Society for Neuroscience, Abstract No. 1852.

Svensson, L.T. (1977). Note on the golden section. Scand. J. Psychol., 18, 79 -80.

Talbot, S.A., and Marshall, W.H. (1941). Physiological studies on neural mechanisms of visual localization and discrimination. Amer. J. Ophthal., 24, 1255 -1263.

Thatcher, R.W. and John, E.R. (1977). "Functional Neuroscience, Vol I.: Foundations of Cognitive Processes", L. Erlbaum Assoc., N.J.

Thatcher, R.W. (1977). On the neural representation of experience and time. In: G. Haydu (Ed.), Experience Forms: Their Cultural and Individual Place and Function, Mouton Press, Amsterdam.

Thatcher, R.W. (1997). Neural coherence and the content of consciousness. Consciousness and cognition, 6: 42-49.

Thatcher, R.W., North, D., and Biver, C. (2008). Intelligence and EEG phase reset: A two-compartmental model of phase shift and lock, NeuroImage, 42(4): 1639-1653.

Thatcher, R.W., North, D., and Biver, C. (2009). Self organized criticality and the development of EEG phase reset. Human Brain Mapp., 30(2): 553-574.

Thatcher, R.W. (2016). "Handbook of Electroencephalography and EEG Biofeedback". Anipublishing, St. Petersburg, Fl.

Thatcher, R.W., Palmero-Soler, E., North, D., and Biver, C. (2016). Intelligence and EEG measures of information flow: Efficiency and Homeostatic Neuroplasticity. Sci. Rep. 6, 38890; doi: 10.1038/srep38890

Thompson, D'Arcy. (1961). "On Growth and Form", Cambridge: University Press.

Tootle, R.B., Silverman, M.S., Switkes and De Valois, R.L. (1982). Deoxyglucose analysis of retinotopic organization in primate striate cortex. Science, 218, 902 -904.

Trentini, B. (2016). "Philosophical Aesthetics and Neuroaesthetics: A Common Future?", *Aesthetics and Neurosciences*, Zoï Kapoula and Marie Vernet (Eds), Springer, chapt. 7.

Tsukiura, T. and Cabeza, R. (2011) Shared brain activity for aesthetic and moral judgments: Implications for the Beauty-is-Good stereotype. Soc Cogn Affect Neurosci 6(1):138–148.

Turner, F., & Pöppel, E. (1988). Metered poetry, the brain, and time. In I. Rentschler, B. Herzberger, & D. Epstein (Eds.), Beauty and the brain: Biological aspects of aesthetics (pp. 71–90). Basel: Birkhäuser Verlag.

Vedder, A., Smigielski, L., Gutyrchik, E., Bao, Y., Blautzik, J., Pöppel, E., . . . Russell, E. (2015). Neurofunctional correlates of environmental cognition: An fMRI study with images from episodic memory. PLoS ONE, 10(4), e0122470. doi:10.1371/journal.pone.0122470

Von Bekesy, G. (1960). "Experiments in Hearing", McGraw-Hill, New York, 1960.

Weiman, C.F. and Chaiken, G. (1979). Logarithmic spiral grids for image processing and display. Comparative Graphics and Image Processing, 11, 197 -226.

Werner, G., and Whitsel, B.C. (1973). Functional organization of the somatosensory cortex. In: R. Iggo (Ed.), Handbook of Sensory Physiology Vol. II, pp. 621-700, Berlin-Heidelberg-New York: Springer.

Werner, G. and Whitsel, B.C. (1968). Topology of the body representation in somatosensory area S-I of primates. J. Neurophysiology, 31, 856 -869.

Wiener, N. (1948). Cybernetics, Cambridge Univ. Press, New York.

Woolsey, C.N., Marshall, W.H. and Bard, P. (1942). Representation of cutaneous tactile sensibility in the cerebral cortex of the monkey as indicated by evoked potentials. Bull. Johns Hopkins Hosp., 70, 399 -441.

Varela F, Lachaux J-P, Rodriguez, E., Martinerie, J. (2001) The brainweb: Phase synchronization and large-scale integration. Nat Rev Neurosci 2(4):229–239.

Vartanian, O and Goel, V. (2004) Neuroanatomical correlates of aesthetic preference for paintings. Neuroreport 15(5):893–897.

Vessel, E.A., Starr, G.G., Rubin N (2012) The brain on art: Intense aesthetic experience activates the default mode network. Front Hum Neurosci 6:66.

Zaidel, D.W. and Nadal, M. (2011) Brain intersections of aesthetics and morals: Perspectives from biology, neuroscience, and evolution. Perspect Biol Med 54(3):367–380.

INDEX

S

T

U

V

W